Space, Time and the Construction of Identity

Linguistic Insights

Studies in Language and Communication

Edited by Maurizio Gotti,
University of Bergamo

Volume 166

PETER LANG
Bern · Berlin · Bruxelles · Frankfurt am Main · New York · Oxford · Wien

Rita Salvi & Janet Bowker (eds)

Space, Time and the Construction of Identity

Discursive Indexicality in Cultural, Institutional and Professional Fields

PETER LANG
Bern · Berlin · Bruxelles · Frankfurt am Main · New York · Oxford · Wien

Bibliographic information published by die Deutsche Nationalbibliothek
Die Deutsche Nationalbibliothek lists this publication in the Deutsche Nationalbibliografie; detailed bibliographic data is available on the Internet at ‹http://dnb.d-nb.de›.

British Library Cataloguing-in-Publication Data: A catalogue record for this book is available from The British Library, Great Britain

Library of Congress Cataloging-in-Publication Data

Space, time and the construction of identity : discursive indexicality in cultural, institutional and professional fields / Rita Salvi & Janet Bowker (eds).
pages cm. – (Linguistic Insights : Studies In Language And Communication, 1424-8689 ; v. 166)
Includes bibliographical references.
ISBN 978-3-0343-1254-7
1. Space and time in language. 2. Identity (Psychology) 3. Language and culture. 4. Sociolinguistics. I. Salvi, Rita. II. Bowker, Janet.
P37.5.S65S63 2013
420.1'9–dc23

2013015536

ISSN 1424-8689 pb. ISSN 2235-6371 eBook
ISBN 978-3-0343-1254-7 pb. ISBN 978-3-0351-0462-2 eBook

Hochfeldstrasse 32, CH-3012 Bern, Switzerland
info@peterlang.com, www.peterlang.com, www.peterlang.net

Contents

Part 2: Corporate and Organizational Discourse

RITA SALVI / JANET BOWKER

Preface

This volume is one of the scientific products resulting from a PRIN project 2008, which gathered the Universities of Modena-Reggio Emilia, Bergamo, Florence, Naples 'L'Orientale' and Rome "Sapienza". The research groups were privileged to have prof. Marina Bondi as national coordinator. The general title of the project was "Discourse identities and sense of space: from key-words to phraseology". Fundamentally, the project aimed at a methodological rethinking about the notion of 'keywords' in relation to phraseology, the influence of the cognitive factors of time and space in the construction of identity, and the integration of discourse analysis with corpus linguistics. The studies collected in this volume, therefore, tend to further illuminate the nature of communication – corporate, institutional, professional and academic – in which the construal, maintenance and repairing of identity develop in the course of time and in different locations or mental spaces.

At the broadest level, it was hypothesized that the various genres, subgenres and variations in contextual use (corporate and institutional websites, internal corporate communications, academic lectures and cultural encounters) would share a great deal in their modes of realization of their major communicative purposes, namely those of creating persuasive and coherent textualizations to project and promote a series of positions and behaviours which pertain to the field of different specialized languages. However, at a micro-level, investigations into the differing contexts of use reveal a range of specific linguistic features as well as pragmatic patterns and strategies employed in both the externally-directed expression and referencing of time and spatial relationships in language, as well as the obverse, the functioning of linguistic indexicality in the internalized shaping of perceptions and cognition ranged along these fundamental experiential dimensions.

The analytical routes used by the authors of this volume to uncover these correlations are varied, using differing theoretical models and empirical approaches: quantitative and qualitative methods assume changing centrality in the chapters, a number of descriptive linguistic frameworks are also pertinent, but a common thread lies in the central role of co-constituent creative negotiation and participant interaction in the construal of identity. In this process the interlocutors' identity is established and reinforced in relation to a sense of place, of belonging to particular social and professional groups, these being characterized by differentiating sets of values, beliefs, norms and practices. The descriptions, then, focus on how personal, professional and corporate identities are shaped through changing social, cultural and ideological perspectives in different temporal and spatial positioning.

In sum, despite the rather daunting extent and abstraction of the categories of description central to this volume – time, space, identity and their relationship to discourse and language – hopefully the specialized, empirical studies contained here shed valuable insights into this vast and challenging area. The findings reported may also be of interest beyond any narrowly defined academic sphere, incorporating relevance for educational and professional practitioners alike.

MARINA BONDI

Introduction

"Another book on identity needs some justification" claims Ken Hyland in the preface of his recent book on disciplinary identities in academic writing (2012: ix). And just restricting the focus to discursive identity or even to the interplay between cultural, institutional and professional identities certainly does not in itself provide a strong rationale. The motivation for this collection of papers lies rather in the need to explore the role played by indexicality – and by the representation of space in particular – in the discursive construction of identity. This volume and the research project from which it develops – a national project on *Discourse Identities and Sense of Space: from Keywords to Phraseology* – investigate how cultural, institutional and professional identities are constructed linguistically through a corpus-informed investigation of discourse. Its distinctive perspective is thus methodological as well as thematic. Combining the tools of corpus linguistics and discourse analysis, the book focuses on indexicality and on the representation of space-time in discourse as important tools of identity construction.

Discourse identity has become an increasingly popular research topic in recent years (De Fina *et al.* eds. 2006; Benwell/Stokoe 2006; Llamas/Watt 2010), across disciplines and analytical approaches, including critical discourse analysis, conversation analysis and 'membership categorization analysis' (Antaki/Widdicombe 1998). Goffman (1967) had already developed an account of the self as social construction bridging the theoretical and analytical divide between self and society. In discourse approaches, participants are seen to draw upon multiple identities in a given communicative context and identities are dynamically negotiated as discourse unfolds (De Fina *et al.* 2006). This can be indexed linguistically through labels, code choices, lexical choices, stances, styles or different structures, which can in turn be related to macro social categories as ethnic group, social class, nationality or gender (e.g. Gumperz 1982).

Indexicality involves the features of language that refer directly to the context in which an utterance takes place: expressions of personal, temporal and locational deixis (expressions such as *we*, *today, this report, here,* and *you*) are typically associated with different meanings (or referents) on different occasions. The notions of person, space and time in general are inextricably linked together and together contribute to establishing a point of view in discourse. By using locatives, personal pronouns and time expressions language users do not only point to their roles as discourse producers and recipients but also to their location in time and space and to their relationship to others (present or absent) (De Fina *et al.* 2006).

Recent socio-anthropological approaches extend to non-literary narratives the bakhtinian notion of chronotope – "the intrinsic connectedness of temporal and spatial relationships that are artistically expressed in literature" (Bakhtin 1981: 84) – and expand its scope. Agha (2007: 320) talks of "'cultural chronotopes', namely depictions of place-time-and-personhood to which social interactants orient when they engage each other through discursive signs of any kind": talking of West and East Germany, for example, would imply referring to a specific time and place, and would normally 'position' the speaker/writer. Starting from similar assumptions, Schiffrin (2009: 243), points at a range of "textual features by which a speaker can construct both personal and place identity".

First and second person pronouns play an important role in constructing social identities because they refer to the interpersonal relations of the discourse participants, but specialized lexis is also used in institutional discourse to display expertise and thus distinguish professional identity in the eyes of lay-person (Drew/Heritage 1992). Certain lexical items can act as 'cultural keywords' in various types of discourse, thus providing interesting insights into the social and personal identities of those who use them, as well as the values and ideals of their discourse community (Williams 1976; Stubbs 2001; Baker 2006). Lexical choice can be seen to reflect the value-system of the speaker and the discourse community and thus act as identity markers.

If the context of a word reflects its underlying value-system, lexical selection shapes discursive identity.

Individuals characterize themselves, or are characterized by others, in terms of ethnicity, gender, profession, social status etc. The obvious association of identity with culture can refer to both the 'large culture', associated with ethnic, national, international groups, and the 'small cultures', associated with activities within a given social group (Holliday 1999). If culture is a repertoire of the shared meanings of a community, the notion can easily extend to disciplinary academic and professional communities. These can be seen as 'communities of practice' (Wenger 1998), i.e. communities defined by co-participation in specific activities and by the kernel of key concepts that characterize their discourse: ways of doing things, ways of talking, beliefs, values, power relations, in short practices. The notion is widely used in academic discourse studies (Hyland /Bondi eds. 2006; Fløttum *et al.* eds. 2007) and can be used for professional discourse too.

Space plays a special role in identity construction, both in cultural and social/institutional/professional terms. Identity helps people 'locate' themselves in their social worlds. By helping to define where they belong and where they do not belong in relation to others, it gives them a sense of 'place' (Scollon/Scollon 2003). The strong connection between space and identity construction has been the central concern of cultural studies (Bhabha 1994) and of works on space, place, and cultural heritage (Malpas 1999). Likewise, space and spatial deixis have been a central concern of cognitive approaches to language and to the interrelation of language, mind and culture (Levinson 2003; Levinson/Wilkins 2006). Discourse studies, however, have only paid little attention to the issue, with the exception of works on description as a text type or part genre (Hamon 1981; Adam 1993; Radighieri/Tucker eds. 2009; Caballero 2005; Gerbig 2008, 2010), with an emphasis on point of view, rather than identity. In applied discourse studies, even when attention is paid to commodified spatial and virtual identities (Benwell/Stokoe 2006), language analysis does not always match professional interest: corporate identity has long been the object of interest in organizational studies (e.g. Balmer/ Greyser 2003), often in relation to the country-of-origin-effect (see Pharr 2005 for a literature review), but analysis of its discursive construction is much more recent (e.g. Garzone/Evangelisti Allori 2010).

All this seems to us to provide ample motivation for the present volume's focus on how indexicals and forms of locational self-representation are substantiated in multiple identity construction processes characterizing professional and institutional practices.

The book also brings together different strands of linguistic investigation, i.e. discourse analysis, with the models of discourse identity mentioned above, and corpus analysis, with the notion of the 'extended unit of meaning' (Sinclair 1996) forming the basis of a phraseological approach to discourse. The phraseological focus links to recent work on keywords (Scott/ Bondi eds. 2010) and extends the exploration to syntactic and semantic patterns. The combination of keywords, 'concgrams' (Cheng *et al.* 2006) with their attention to patterns of inter-collocation of collocates, key semantic domains (Rayson 2008) and semantic sequences (Hunston 2008) is shown to allow both macro-analysis (studying the features of whole texts or varieties of language) and micro-analysis (focusing on the use of a particular language feature). The cultural significance of a lexical item can be brought to the fore focusing on its collocations and 'semantic preferences' (i.e. its ability to co-occur regularly with words belonging to a specific semantic field, Sinclair 1996) or on its pragmatic implications and preferred communicative contexts.

The volume is divided into two parts, focusing respectively on institutional and cultural discourse (Part 1) and on corporate and organizational discourse (Part 2).

Part 1 opens with a discussion of identity by RITA SALVI. Starting from the notion of identity as a cognitive phenomenon, essential to perform any human behaviour in a physical environment (space) and within a phenomenological context (time), the chapter focuses on how this 'private' aspect of cognition and experience is constituted in the domain of public discourse. This is particularly important when looking at institutional identity, which is so obviously contingent on local conditions and different interactional contexts. Space and time are seen as an integral part of discursive patterns, as shown in the chapter through a case study of professional and institutional identity in the genre of economics lectures.

Personal pronouns – and other markers of person indexicality – are seen to play a crucial role in the construction of discursive identities, as they establish the point of view of discourse and the *persona* attributed to both addresser and addressee. The study by GIULIANA DIANI looks at academic identity contrastively, in English and Italian university lectures. She looks in particular at the construction of a voice for lecturers and their positioning in relation to their knowledge claims, their community, and their students. The chapter presents the results of a contrastive study of first person pronouns in English and Italian lectures in the field of economics.

The role played by discourse in the representation of the European Union as a political and social actor is analysed by GIUDITTA CALIENDO in a corpus of informative and popularizing booklets which the EU has published in the EUROPA website over approximately ten years. The analysis highlights the way in which rhetorical and discursive strategies, reflecting the identity values foregrounded, have undergone significant changes over the last decade, in line with a wide series of official action plans calling for a drastic reform of EU communication and information strategies.

CHIARA PROSPERI PORTA looks at the shaping of identities in the annual reports of European central banks in the context of the financial crisis and recession. The study considers the Annual Reports of three European Member States, comparing them with the same type of documents issued by the European Central Bank over the period that goes from 2008 to 2011. The analysis shows how the common problems currently affecting Europe are being differently perceived and shaped at a national or supranational level.

The study of LAURA FERRAROTTI addresses the promotion of institutional identity in Italian university websites. Based on a corpus of English versions of Italian university websites, the study concentrates on how personal deictics *we/you/our/your* are used in introductory sections of university websites. In line with the general trend of marketization of public institutions, universities are shown to strategically address their audience either by including or excluding them, in order to project an authoritative image.

The chapter by GIUSEPPE BALIRANO and MARIA CRISTINA NISCO rounds up the first part of the volume, centring on hybrid cultural identities. Films representing the process of identity construction of South-Asian immigrants in Western countries are shown to characterize the new diasporic identities through specific labels, code choices, lexical choices, stances, styles or linguistic structures. The analysis suggests that audiovisual translation necessarily calls for new strategies to render acts of identity, markers of resistance, abrogation or appropriation in order to re-locate English in the target language.

The main focus of Part 2 is on the construction of corporate or organizational identity in external communication, notably in disclosure genres (financial and corporate social responsibility reports) and websites. Corporate identity is the self-presentation of an organisation, how it views itself and how it wants to be seen. This includes its 'mission and vision': its business strategy, its organizational culture and structure and especially its values. Building and maintaining a corporate identity and reputation has become a keystone in the strategic communication of companies. Disclosure documents have become particularly important in building a positive image and maintaining a reputation. Annual reports are crucial in creating a strong and trust-based relationship with stakeholders, establishing the company's positive identity and building or preserving a reputation, especially in moments of crisis. A discussion of corporate identity inevitably brings in what is perhaps the best known tools for self-promotion, i.e. the corporate website. The Internet has become the preferred channel for disseminating information in general, but its role in the corporate context cannot be overestimated: it is an extremely efficient tool for distributing company communications at a low cost.

JANET BOWKER looks at how identity is constructed in the multimodal, virtual professional space of corporate webcasting events. Virtual spaces are converted – in and through language – into purposeful places. She investigates the relationship between channels and embodied discourse and the processes of 'grounding', the anchoring of 'unboundedness' in the *here and now*. People orient themselves and their discourse to the medium's relevant features, while dynamically reconstructing the space itself, at the same time.

The case study presented by CHRISTINA SAMSON is to be set against the background of recent unethical financial reporting practices, and of a general debate as to whether reporting narratives deliberately mislead investors or are effectively useful. The study looks at lexical-phraseological representations of identity in a specialised corpus of performance and risk reviews extracted from the corporate annual reports of the Italian Stock market, in order to investigate the most frequent lexical items and phraseology encoding corporate identity and reputation.

ERSILIA INCELLI provides an example of how phraseological analysis can help trace the ways in which reputation is managed in a situation of conflict. The chapter addresses the dispute that emerged between the diamond industry and its opponents in the wake of growing social concern for the role of 'conflict diamonds' in Africa (1997-2011). With its image heavily damaged and its legitimacy undermined, the industry adopted new discourse strategies in order to reconstruct a positive image in the eyes of its stakeholders, by framing responsibility to the environment and community.

As also shown by DONATELLA MALAVASI, websites appear to act as a virtual shop window, having the overt intent to market a product/ service and attract clients, but they also promote brand identification, gain support of the firm's internal staff, and enhance investors' trust by constructing of a sound corporate identity and projecting a commendable image. Malavasi's study centres on local and global distinctive traits of the 'Made in Italy', investigating how identity is enacted by a selection of Italian firms in three key industries, i.e. Fashion, Food and Furniture. Particular attention is paid to the representation of space in the websites of the firms studied and the role the country of origin plays in their self-appraisal.

Along similar lines, GILLIAN MANSFIELD and FRANCA POPPI look at the country-of-origin effect of well-known Italian products (from food and drink to ceramic tiles) in global advertising. The aim is to project a clearly identifiable and relevant cultural self to the foreign market by maintaining essential lexis in the original language or by highlighting personal commitment to values. With reference to a corpus of Italian company websites and a corpus of multimodal ads, the chapter reveals how the Italian-ness of both the company and its product(s) is conveyed verbally and non-verbally. The analysis shows that products from extreme ends of the market sector share an interest in establishing and reinforcing identity in relation to a sense of place and of belonging to a particular national and/or social group, through values conveyed implicitly by the association.

JUDITH TURNBULL rounds up the section by looking at CSR (Corporate Social Responsibility) communication and its role in building, enhancing and defending reputation. Her focus is on the status of a multinational operating all over the world with its concomitant reputation, where time and space act as key coordinates in the construction of corporate identity and reputation. The case study presented is that of *Unilever*, an Anglo-Dutch corporation, which produces some of the world's leading consumer products in food and beverages, personal and home hygiene. Through a close analysis of a section of *Frequently Asked Questions*, tackling controversial aspects of multinational business, the study shows how *Unilever* positions itself in relation to different stakeholders.

The volume draws attention to the interconnectedness of person, time and space in identity construction as we create meanings in professional interaction. The research group sets out to offer an overview of different aspects and practices of our participation in professional communities, highlighting how these are influenced and shaped by the performance of identity in discourse.

References

Adam, Jean-Michel 1993. *La Description*. Paris: PUF.

Agha, Asif 2007. Recombinant Selves in Mass Mediated Space-time, *Language & Communication* 27, 320-335.

Antaki, Charles / Widdicombe, Susan 1998. *Identities in Talk*. London: Sage.

Baker, Paul 2006. *Using Corpora in Discourse Analysis*. London: Continuum.

Bakhtin, Michail 1981. *The Dialogic Imagination: Four Essays*, Michael Holquist (ed). Austin/London: University of Texas Press.

Balmer, John M. T. / Greyser, Stephen A. (eds) 2003. *Revealing the Corporation: Perspectives on Identity, Image, Reputation, Cor-*

porate Branding, and Corporate-level Marketing. London: Routledge.

Benwell, Bethan / Stokoe, Elizabeth 2006. *Discourse and Identity*. Edinburgh: EUP.

Bhabha, Homi 1994. *The Location of Culture*. London: Routledge.

Caballero, Rosario 2005. *Re-Viewing Space*. Berlin: de Gruyter.

Cheng, Winnie / Greaves, Chris / Warren, Martin 2006. From n-gram to skipgram to concgram. *International Journal of Corpus Linguistics*. 11/4, 411-433.

De Fina, Anna / Schiffrin, Deborah / Bamberg, Michael (eds) 2006. *Discourse and Identity*. Cambridge: CUP.

Drew, Paul / Heritage, John (eds) 1992. *Talk at work: Interaction in institutional settings*. Cambridge: Cambridge University Press.

Garzone, Giuliana / Evangelisti Allori, Paola (eds) 2010. *Discourse, Identities and Genres in Corporate Communication*. Bern: Peter Lang.

Fløttum, Kjersti / Dahl, Trine / Kinn, Torodd / Gjesdal, Anje Müller / Vold, Eva Thue 2007. Cultural identities and academic voices. In Fløttum, Kjersti (ed.) *Language and Discipline Perspectives on Academic Discourse*. Newcastle: Cambridge Scholars, 14-39.

Gerbig, Andrea 2008. Travelogues in time and space. In Gerbig, Andrea / Mason, Oliver (eds) *Language, People, Numbers*. Amsterdam: Rodopi, 157-173.

Gerbig, Andrea 2010. Key words and key phrases in a corpus of travel writing. In Scott, Mike / Bondi, Marina (eds) *Keyness in texts*. Amsterdam: Benjamins. 147-168.

Goffman, Erving 1967. *Interactional Rituals. Essays on Face-to-face Behavior*. Anchor Books, New York.

Gumperz, John (ed.) 1982. *Language and Social Identity*. Cambridge: CUP.

Hamon, Philippe 1981. *Du Descriptif*. Paris: Hachette.

Holliday, Adrian 1999. Small Cultures. *Applied Linguistics*. 20/2, 237-264.

Hunston, Susan 2008. Starting with the small words: Patterns, lexis and semantic sequences. *International Journal of Corpus Linguistics*. 13/3, 271-295

Hyland, Ken 2012. *Disciplinary Identities. Individuality and Community in Academic Discourse.* Cambridge: CUP.
Hyland, Ken / Bondi, Marina (eds) 2006. *Academic Discourse across Disciplines*. Bern: Peter Lang.
Levinson, Stephen 2003. *Space in Language and Cognition*. Cambridge: CUP.
Levinson, Stephen / Wilkins, David 2006. *Grammar of Space. Explorations in Cognitive Diversity.* Cambridge: CUP.
Llamas, Carmen / Watt, Dominic (eds) 2010. *Language and Identities*. Edinburgh: EUP.
Malpas, Jeff 1999. *Place and Experience. A Philosophical Topography.* Cambridge: CUP
Pharr, Julie M. 2005. Synthesizing country-of-origin research from the last decade: is the concept still salient in an era of global brands? *Journal of Marketing Theory and Practice*. 34-45.
Radighieri, Sara / Tucker, Paul (eds) 2009. *Point of View: Description and Evaluation across Discourses*. Roma: Officina.
Rayson, Paul 2008. From key-words to key semantic domains. *International Journal of Corpus Linguistics*. 13(4): 519-549.
Schiffrin, Deborah 2009. Crossing Boundaries: the Nexus of Time, Space, Person, and Place in Narrative. *Language in Society* 38, 421-445.
Scollon, Ron / Scollon, Suzie Wong 2003. *Discourses in Place*. London: Routledge.
Scott, Mike / Bondi, Marina (eds) 2010. *Keyness in Text.* Amsterdam: Benjamins.
Sinclair, John 1996. The search for units of meaning. *Textus*. 9/1, 75-106.
Stubbs, Michael 2001. *Words and Phrases*. Oxford: Blackwell.
Wenger, Etienne 1998. *Communities of Practice. Learning, Meaning and Identity.* Cambridge: CUP.
Williams, Raymond 1976/83. *Keywords: A Vocabulary of Culture and Society*. London: Fontana.

Part 1
Institutional and Cultural Discourse

RITA SALVI

Space and Time in the Construction of Identity in Institutional Discourse

> [...] despite the standardising pressure of cultural and language-related factors, academic communication remains in many ways a highly personal affair, with active participation in a disciplinary community requiring a multi-dimensional discourse that combines the professional, institutional, social and individual identities of its members. (Gotti 2012: 9)

1. Introduction

The construction of identity has been a central topic in the last years (Bucholtz/Hall 2005; Edwards 2006; Fairclough 2003; Salvi 2007), dealt with from several interdisciplinary approaches. Most present day theories, however, agree in considering identity as a cognitive phenomenon, essential to perform any human behaviour in a physical environment (space) and within a phenomenological context (time). The 'private' aspect of cognition and experience (the *inner* self, according to Benwell/Stokoe 2010) has been matched and integrated with an 'outer' self, revealed in the domain of public discourse, in which identity is shown. More than that, Benwell and Stokoe state that "rather than being *reflected* in discourse, identity is actively, ongoingly, dynamically *constituted* in discourse" (Benwell/Stokoe 2010: 4, italics in the original).

Therefore, parallel to 'personal identity', 'social identity' has been studied (Social Identity Theory, SIT, Tajfel/Turner 1986) to define the individual's stance within a group, its outward projection and the extent of his/her identification with the group. This can be the case of experts operating in institutions: their sense of belonging affects the use of the language adopted to transfer information which will become the shared knowledge in the discourse community. Although institutional discourse requires stable communicative practices, institutional identity cannot be considered a 'fixed' entity, as it is contingent on local conditions and different interactional contexts.

Space and time, which are also cognitive categories, so much so that individuals give a personal interpretation to both, have to be considered an integral part of discursive patterns with reference to institutional communication as well as to the construction of identity (personal and social).

This chapter analyses spatial and time references in the construal of identity in the genre of lectures on economics.

2. A theoretical framework for space, time and institutional identity

The objective of this research is to demonstrate if and to what extent the categories of space and time contribute to the construction of an institutional identity. I intend here to follow the geographer Edward Soja's theory (2001) which overcomes a binary conceptualization of space. He recognizes a dichotomy between the 'perceived' space (which consists mainly of concrete spatial forms, things that can be empirically mapped) and the 'conceived' space (a spatial variety adopted on some occasions and in specific circumstances; it is constructed in mental forms and expressed in systems of signs and symbols). He also argues that space is a culturally constructed entity which can be formed and changed, accepted or rejected. Therefore, he introduces the concept of a 'third space' which is a 'lived space' beyond physical ('perceived') form or mental ('conceived') construct. Lived space embodies the lifeworld of events, experiences and emotions. It draws upon both the material (perceived) and the mental (conceived) spaces, but extends beyond them in scope, substance and meaning. This type of space is created by the effects of a changing culture and is characterized by simultaneity and transience. In this sense 'spatiality', which is an essential aspect of human life and description of the world, is a 'socially produced space'.

Soja's geographical theory can be integrated by Fauconnier's distinction (1997) between 'base space' and 'mental space', which will help us to approach language analysis more directly. In particular, he considers the present reality as a 'base space' (the mutually known world of the interlocutors, Croft/Cruse 2010: 33). 'Mental space' is instead a cognitive structure, consisting of complex conceptual interconnected networks, constructed in the course of speaking or thinking. Fauconnier maintains that mental space is context-dependent and includes information about specific situations. The bridge between the two types of space is built by roles (the position of a single individual at one time and of different individuals over time) and values (abstract concepts or theoretical assumptions adopted to explain reality): both roles and values are factors which can be analysed within institutions.

In line with Fauconnier's theory, 'space builders' can be detected, that is "a grammatical expression that either opens up a new space or shifts focus to an existing space" (1997: 40). Together with grammatical expressions, Golshaie argues that names and descriptions "either set up new elements [of space] or point to existing elements in the discourse construction" (2011: 46). With reference to space, Golshaie also claims that "conceptual networks or domains are formed in the working memory and are expanded as the process of thinking or conceptualization continues" (2011: 45).

As far as time is concerned, scholars agree that time is an abstract domain and only some elements of time are present in our experience with the world. This experience also confirms that each moment in time only happens once and that time admits only unidirectional changes. For the purpose of the present research the psycholinguistic ego-moving schema ("in which the 'ego' or the observer's context progresses along the time-line toward the future", Boroditsky 2000: 5) seems suitable to describe the speaker's position and stance, underpinning his role in institutions.

Traugott (1978) divided the time domain into three components, namely tense, aspect and sequencing. In the case of this study, sequencing is the most relevant aspect, as it is a system in which events are ordered with respect to each other and to the speaker. Galton states

> Time and space go together: it is hard to think about either without thinking about the other. They are our basic location frameworks [...] To locate an event we must specify both a place and a time; if either is missing, our knowledge of the event is incomplete. (Galton 2011: 695)

As we will see, this is an important aspect in academic discourse, particularly in the field of economics, in which each event is part of a sequence, or an element of a cause/effect chain. In any case, the high level of connection between space and time is the reason for which the language of space is adopted to describe time, as the abstract domain of time can be organized in terms of the observable domain of space.

As Lehrer argues, "the spatial meaning of many prepositions carries over to the temporal domain in predictable ways" and the two domains even utilize the same words (1990: 220):

Space	Time
at the corner	*at* noon
from here *to* there	*from* two o'clock *to* four o' clock
through the tunnel	*through* the night
he stood *before* the house	it happened *before* evening
he was running *ahead* of me	he arrived *ahead* of me

Table 1. Space-time correspondences in language.

An analysis of institutional identity cannot disregard the language used to present particular events in specific contexts, because

> Institutions are abstract structures, which change historically, but typically exist over long periods of time. They nevertheless depend on their constituent speech events, which in turn depend on speech acts which last for only seconds. The problem is to relate things of different scales in time and space, and the main ontological nightmare involves the dualism of agency and structure. (Stubbs 2010: 32-33)

Given that "institutions can comprise any powerful group and are not restricted to material locations" (Benwell/Stokoe 2010: 88) and considering that "scholarly discourse is not a single uniform and monolithic entity [...] instead, it has come to be regarded as an outcome of a multitude of practices and strategies, where argument and engagement are crafted within communities" (Hyland/Bondi 2006: 7), lectures will be analysed as the typical discourse of academic institutions, which involves a variety of topics discussed and explained in different periods and locations, epitomizing both personal and institutional identity.

3. Corpus and methodology

Lectures can be said to represent a prestigious form of institutional discourse. Far from being a static genre, lectures result in typified rhetorical actions inherent to social and cultural situations in an evolutionary perspective, requiring flexibility and innovation on the one hand, and concerned with the audience's expectations on the other (Bhatia 2008). They fully show the interactive nature of academic discourse, which in turn involves communication between experts, and between experts and students.

An analysis of lectures on economics has already been the object of a recent study (Salvi 2012), for which a corpus was built, consisting of two sub-corpora. They include lectures given by John K. Galbraith (1908-2006) and Joseph E. Stiglitz (1943), two scholars in the field of economics, whose personal prestige is an incarnation of "the epistemic conventions of the discipline" (Hyland 2006: 21). Specifically, the study analysed the five brief lectures that John K. Galbraith gave in 1961, when he was the United States ambassador in India. The lectures were given at the Universities of Madras, Calcutta, Bombay, Rajasthan and at the Indian Institute of Public Administration of New Delhi. These lectures were published in a volume (1962). The text results in a small corpus of 14,292 words.

A comparison was made with a recent and rather ample lecture ('Making Globalization Work') that Joseph E. Stiglitz, Nobel Laureate 2001 in Economics, gave in Dublin on August 30 2006, on the occasion of the Geary Lecture, an annual event organized in honour of Prof. Roy C. Geary (1896-1963), the prestigious Irish economist. The lecture was published in 2008. The corpus consists of 9,509 words.

The interesting results in terms of *aboutness* and *keyness* derived from the comparison of these lectures, delivered at almost fifty years' distance of time, have encouraged the enlargement of the corpus to include the Nobel Prize Lecture (1987), 'Growth Theory and After', given by another outstanding economist, Robert Solow (born in 1924). As a result, the corpus analysed in the present research consists of:

- Galbraith 1962 – India 14,292 total words – 2,604 unique words
- Solow 1987 – Sweden 8,697 total words – 1,940 unique words
- Stiglitz 2008 – Ireland 9,509 total words – 1,726 unique words

The different topics discussed by the speakers, as well as the different periods and locations of the communicative events, allow us to better identify the application of the categories of space and time as systems of meaning-making aimed at the construction of personal and social identity.

The three sub-corpora are comparable in that they treat the same discipline and are the result of an oral communication reviewed for a written edition. They are also comparable on account of the number of items, although in the present research reference to numerical data is made only to support discourse analysis (DA) and critical discourse analysis (CDA). They are, in this case, the more suitable approaches in interpreting lectures which are an example of the way people interact in social situations and reflect existing macro-social forces. The research is also an attempt to show how the same material, namely the same corpus, can be 'read' and analysed from different points of view.

Analysing the language used in these contexts is of fundamental importance, as "It is the primary work of language to make all those 'other' phenomena [experience, reality, feelings, or knowledge] accountable" (Edwards 2006: 42).

4. The language of lectures in a spatial/temporal perspective

Although not a homogeneous genre, lectures have maintained a macrostructure generally constituted by a certain number of expositions within which a 'focal episode', 'various developmental episodes' and a 'closing episode' are collocated (Bhatia 2004: 46). They aim at enlarging the consensus of the audience about the theories discussed and, based on the speaker's authority as the linchpin of the communicative event, they intend to persuade and stimulate complicit cooperation for prospective pragmatic decisions. Furthermore, lectures on economic topics, usually published in English, are intended for a supranational scientific community. This involves a high degree of shared knowledge as well as an interwoven structure of contents and contexts, which corresponds to a deep level of intertextuality as described in Fairclough's framework: "basically the property texts have of being full of snatches of other texts, which may be explicitly demarcated or merged in, and which the text may assimilate, contradict, ironically echo, and so forth" (1992: 84). This is of particular interest in the field of economics, in which discourse and argumentation are based on a hypothesis/thesis pattern, a problem/solution model or a chain of cause/effect relations, and include several ramifications in the discipline which influence rhetorical choices and discursive strategies (Salvi 2011).

4.1. Lexical indexicality

The relationship between the topic (*aboutness*) and the selection of words (*keyness*) in economic lectures has been documented in the research mentioned above (Salvi 2012): the quantitative analysis of Galbraith's and Stiglitz's sub-corpora has already given evidence of the extent to which lexical choice is dependent on content. The term *development*, for example, is actually the first in the frequency list for Galbraith (115 occurrences, 0.8%), a word which appears only once in

Stiglitz. *Globalization* is the core matter in Stiglitz (45 occurrences, 0.4%), which justifies the extensive use of the words *countries* (66 - 0.69%) and *country* (28 - 0.29%). Confirmation of the relationship between keyness and aboutness comes from Solow's sub-corpus where *growth* appears 115 times (1.32%).

Several words which constantly appear in the whole corpus, such as *planning, process, stage, change(s), goals and achievements*, contain the concept of evolution in time and space positioning. Beyond lexical frequencies, however, the meaning of lectures is highly dependent on contexts and speakers, so space and time references do construct the cultural environment shared by the discourse community. In other words, space and time references allow the participants to understand the topic in the same way. In the three sub-corpora the concept of 'change' is referred to different contexts, expressed with different words such as *process* or *development*, also with unusual comparative forms (*green/greener*), and the use of verbs such as *become* and *get*. The concept of 'change' is also applied to social, political and economic factors as well as to theoretical issues, in a strongly contextualized form which requires different levels of interpretation according to the further approximations implied in each step:

(1) [...] we must recognize that economic development is a process - one that extends in range from new nations of Africa but slightly removed from their tribal structure to the elaborate economic and social apparatus of Western nations. At each stage along this continuum there is an appropriate policy for further advance. What is appropriate at one stage is wrong at another. (Galbraith)

When I read Robert Frost's lines from "The Black Cottage":
'Most of the change we think we see in life
is due to truths being in and out of favor'
it occurred to me at once that they sound altogether too much like economics. (Solow)

There are many other changes going on that are also quite interesting. They are growing broccoli in Greenland. Greenland used to be less green than it is now, but it is becoming greener than it has ever been as a result of global warming. There was a recent article about some people in Alaska buying air conditioners because it is getting so warm in the summer. (Stiglitz)

4.2. Embedded deictic references

For the linguistic analysis, it will soon be evident that the mere detection of deictics in the texts is not sufficient to completely illustrate the dynamics of economic discourse as developed during a lecture. *Now*, for example, a marker of both economic discourse and of lectures as a genre (Salvi 2012), is used as an adverb, meaning 'at present', often determining a contrast with the past which is the real heart of the matter:

(2) The new countries of Asia and Africa are now concerned, as were those of Western Europe in the late eighteenth and early nineteenth centuries, to understand the processes on which progress depends. (Galbraith)
I may not have been as clear then as I am now about the distinction between the two notions of instability. (Solow)
Billion people who had previously been excluded from the global trading system and consequently the global economy are now becoming integrated, partially as a result of globalisation. (Stiglitz)

Now, however, also serves to introduce the speaker's intention, so belongs to the typical use of metadiscourse, as the specific genre requires:

(3) I should like now to comment on this planning. (Galbraith)
But now I have to report something disconcerting. (Solow)
Now, let me switch to a second problem area – global warming. (Stiglitz)

Let us take another example, an apparently simple sentence in Galbraith, where the adverb *here* is used: "The first task here is not to get capital or technicians but to build competent organs of public administration". Does *here* refer to a place, does it limit a space? Not in the sense we generally use it. The reading of a larger part of the paragraph is helpful for a better understanding:

(4) We have said that capital and technical knowledge are the missing elements. But in many of the newer African states national government is still in its beginning stages [...] Under these circumstances investment, whether public or private, is subject to the risks, uncertainties, and eccentricities of poor public administration. It is idle to imagine that good development plans can be cre-

ated or carried out without a good government to do it. [...] The first task here is not to get capital or technicians but to build competent organs of public administration.

Here then includes all the negative conditions which prevent investment and obstruct efficiency. The use of the adjective *newer* implies a comparison with the past, which the audience must be able to perceive. The mention of African states (or, later in the original text, Latin American countries) is not strictly restricted to any locational deictics, but is indicative of the embedded referencing (temporal and spatial) pertinent to the content of academic lectures. The same comment can be made if we consider a sentence from Solow's lecture:

(5) Growth theory did not begin with my articles of 1956 and 1957, and it certainly did not end there. Maybe it began with *The Wealth of Nations*; and probably even Adam Smith had predecessors. More to the point, in the 1950s I was following a trail that had been marked out by Roy Harrod and by Evsey Domar, and also by Arthur Lewis in a slightly different context. Actually I was trying to track down and relieve a certain discomfort that I felt with their work. I shall try to explain what I mean in a few words.

There in this case only marginally refers to the years, let alone a place; rather it focuses on Solow's contributions to the theory, the articles he wrote so long ago and which justify his outstanding position. *There* also opens to a manifest intertextual dimension, generating a contrast between the speaker's approach to the theory and previous researchers. The use of *when* in Stiglitz fulfils the same pragmatic objective:

(6) Economic globalisation has outpaced political globalisation in terms of the change of mindset. Globalisation has meant that we have become more integrated, more interdependent. Greater inter-dependence means that we have to act together; we have, as economists would say, extra responsibilities which mean that we have more need to act co-operatively. However, we do not have the institutions, and in particular the democratic institutions, or the mindsets, to do this effectively. You see this so clearly in the way decisions are made inside the White House – something I saw first-hand. When we were talking about an issue of domestic policy, we always thought of 'what is fair, what is efficient?' However, when it came to international issues, the word 'fairness' was never used. The question was: what is best for the United States? What is best for American companies?

When does not recall a time, but the occasions in which the economic discourse about globalization materialized; the place where it happened is the White House, which obviously entrusts the speaker with personal prestige. Moreover, he wants to involve the scientific community directly ("we [...] as economists") in the conceptual meaning of 'fairness' which needs to be understood at an intertextual level, according to the significance given by the economists in different periods and in different contexts: in our case, 'fairness' is a word which shapes knowledge in the form of layers that have been deposited within a certain space over time (Foucault 1969).

4.3. Space and local places

The first point of analysis is physical space, the places around which lectures are construed.

Given the geographic location of the lectures, it is not surprising that India and Pakistan are mentioned several times in Galbraith's lectures. Galbraith also mentions Africa, Europe, China, Japan and the Soviet Union in order to discuss his topic, economic development, and make comparisons. In Solow's lecture the number of countries mentioned is absolutely irrelevant, as growth theory does not belong to a single nation or state; so we can say once again that the quantitative analysis is not enough to suggest an indication in terms of spatial references to locate the communicative event. The number of countries mentioned in Stiglitz is definitely higher than in the other two subcorpora, which is in line with the topic at issue, globalization. The United States is the country most frequently mentioned in all the lectures, but this does not indicate so much the origin of the speakers as the economic content of the texts: it can be simply confirmed that, in the course of almost fifty years, the United States have maintained its leading role unaltered, even though the world scene has changed.

In lectures in the field of economics, the category of space can be identified in linguistic features other than places. For example, Stiglitz expresses the concept of space in percentages ("in the range of 7 to 8 per cent"); Solow in the movement from one situation to another ("from slow growth to fast growth"; "a small sideward step could lead to eventual disaster"; "ups and downs of the business cycle"); Galbraith relies on the inclusion/exclusion (in/out) relation ("other countries have broken out of the circle"; "if a corporation must be protected in its personality from intrusion by outside authority upon its decisions, [...]").

Indeed, in the lectures examined, the topic is itself a space, in which the speaker intentionally moves his thought. Space and time overlap as the titles all include linguistic elements (Economic Development *in Perspective*, Growth Theory and *After*, *Making* Globalization *Work*) which do not only define a territory of investigation, rather introduce a sequential order of expected events which involve peoples and countries, beyond geographical borders.

4.4. Base and mental spaces

Base spaces in the three sub-corpora can mainly be traced in two types of references: time and theory. As far as time is concerned, it is interesting to note that in the language of economics dates and places, as Galton (2011) suggests, need to be mentioned together, such as in the following sentences:

(7) In five major Indian educational communities [...] in the summer and autumn of 1961 (Galbraith);
In the US between 1959 and 1978 [...] (Solow);
Argentina had a crisis in 2001 (Stiglitz).

The concrete experience of the world economy is marked by dates known within the scientific community, although the way to indicate the time is linguistically different in the three examples. Even when a date is not explicitly indicated:

(8) Not long after the end of World War II (Galbraith);
In the course of preparing this lecture, I came across a recent working paper by Professor Edward N. Wolff (Solow);
Remember during the Industrial Revolution, or the periods of the highest success in the United States (Stiglitz)

time references are the backbone of the presentation, because they contribute to both the development of the topic and the speakers' positioning. The next group of examples, instead, gives the opportunity to reflect about the interpretation of time references:

(9) A few weeks ago President Kennedy announced the first of a series of steps designed to carry a man to the moon (Galbraith);
Today I would put the unsolved problem as follows (Solow);
Just two years before, the country [Moldova] had no debt [...] (Stiglitz).

The meaning of the sentences is highly dependent on the period in which each speech event occurred. *A few weeks ago* in Galbraith is definitely located in the 1960s, and transmits the enthusiastic immediacy of the news to the audience, a perception which can be lost in the course of time. Much more complex is the meaning of *today* in Solow: first of all it refers to the period, not exactly to the day on which the speech occurred. Its meaning is understandable only in the description of Solow's theory, which is in contrast with the Harrod-Domar model (1939 and 1946): the speaker's stance communicated by *today* opens the way to a mental space, in this case evidently constructed and modified as thought and discourse unfold. Also *just two years before* in Stiglitz represents a mental space, and the whole network needs to be reconstructed through the text. The speaker mentions his personal experience in Moldova ("I was in Moldova about four years ago", "When I visited there") as a form of narrative mode used to support his theory. This contributes to define his personal identity in terms of positioning and authority. As the lecture was delivered in 2006, and the speaker visited the country "four years ago", "just two years before" means more or less 2000. The scientific community knows that in that period the transition from a planned to a market economy induced Eastern countries to borrow money, and their debt was in hard currency (German marks and dollars) not in their own currency: the economic background explains the temporal extension and the mentioned place.

So, place and time also define the identity of the institutional environment through shared knowledge: they are connected to each other by various kinds of mappings, in particular analogy mapping (the example of Moldova covers all countries in the same situation) and the affirmation of personal identity coinciding with institutional identity.

Time references are also needed to develop theoretical issues. Indeed, each of the lecturers assumes a precise event, a base space, as a sort of watershed from which the economic topic can be expanded: World War II in Galbraith, the Great Depression in Solow, the fall of the Berlin Wall in Stiglitz. Several layers of intertextuality have offered a large amount of information about them. But they are not episodes, rather a connection of sequential occurrences and consequential relationships with a great impact on economics in the course of time. They can be considered as 'space builders' which are to be detected not only through the identification of linguistic grammar elements (prepositions, adverbs, and so on); they can also be identified in the discourse structure, particularly with reference to Keynes and the neoclassical model, which the three speakers use either to support or discuss the theory. Mental spaces are then activated in the long-term memory of both the speaker and the institutional academic community, because the speaker's intentionality takes on an institutional relevance.

4.5. Personal and institutional identity

The focus in this section is about the ways in which the 'lived space' is negotiated and socially achieved in lectures. For their very nature, lectures present the two basic qualities of the third space, e.g. simultaneity and transience, as they are a communicative event placed in an architectural space where the speaker has a prominent physical position which nonetheless implies a high degree of interaction with the audience, and they last for a limited length of time.

The prominent position and the stance of the speakers is evident in the frequency lists of the sub-corpora, where the personal pronoun *I* shows a high number of occurrences, either as a marker of self-representation or as a means of metadiscursive practice; for example:

(10) I have said that the present diagnosis of the causes of underdevelopment [...] (Galbraith)
This is the sort of amendment that I mentioned in 1956. (Solow)
I want to give a quick description of what I see as the three underlying core problems: trade, financial markets and resources. (Stiglitz)

Far from restricting the speaker's identity, confining his status or limiting his prerogatives, the use of *my* covers the same functions:

(11) Here, in my view, the dangers are greatest of all. (Gabraith)
After this methodological digression, I should remind you of the direction of my main argument. (Solow)
I want to preface my remarks by talking about a question I am often asked – how did I come to write my book entitled Making Globalization Work? (Stiglitz)

The sentences show how the self is fully expressed in public discourse, where it is externalized in the 'knowledge of the world' which has to be shared by all competent members of the institution as 'cultural knowledge' (van Dijk 2003: 90).

The use of *you* is instead limited, very often it refers directly to the participants (as in Solow's example above, "I should remind you"), in line with the purpose of lectures to achieve consensus avoiding divergences or positions of contrast between speaker and audience.

The negotiation of values happens with the use of *we*, which is much more than a marker of positive politeness to create solidarity and empathy among the participants. Actually *we*, occasionally used in the impersonal form for *people*, covers two large areas of inclusiveness, namely the scientific community of the economists, and the socio-political community of American and Western countries.

(12) We – the economists

It would be well were we all to realize that if the pace of less favourably situated countries is slow it is not necessarily because their efforts are less. (Galbraith)
This is the sort of configuration we have come to call "Keynesian". (Solow)

> We discuss the problem of debt in poor countries as if it were just one country or two countries where people do not know how to live within their means. (Stiglitz)

(13) We – the Americans/Western countries

> As a nation we have a healthy respect for money and its uses. (Galbraith)
> My tentative resolution of the dilemma is that we have no choice but to take seriously our own direct observations of the way economic institutions work. (Solow)
> There is absolutely no doubt that we can reduce our emissions enormously without sacrificing our standard of living. (Stiglitz)

Thanks to this use of 'institutional *we*', we can note that – also due to the roles of the speakers and the values they express – simultaneity and transience belong to the surface structure of the lecture: these characteristics are necessary to shape the type and the direction of communication, but it is the common built space (may it be a country, an economic model or a technical decision) which establishes a set of interwoven connections and processes which can be shifted, anchored, or focused on by each speaker, behind the scene of the occasional event, according to his personal experience and professional qualifications. The elaborate web of connections is mainly expressed in terms of values, together with attitudes and ideologies, and the corresponding use of evaluative language ("It is my unhappy feeling", Galbraith; "I think it unlikely", Solow; "I think it is unfair to say", Stiglitz) in utterances which reveal situated use and institutional intelligibility, and not only a personal emotion or an individual mental experience.

5. The third space: roles and values

Having considered the weight of values and ideologies in this type of communicative event, we can focus now on the 'socially produced space' built by each speaker (and the three speakers over time) who transmits knowledge (norms, beliefs and values) within the academic institution throughout his lecture and beyond.

Although it would not have been necessary, none of the speakers is reticent about reaffirming his authority:

(14) When these lectures aroused some interest, not only in India but elsewhere, I was persuaded without great difficulty to rearrange them in this form. (Galbraith)
I know that even as a student I was drawn to the theory of production rather than to the formally almost identical theory of consumer choice. It seemed more down to earth. I know that it occurred to me very early, as a natural-born macroeconomist, [...] (Solow)
I was very pleased that they were reading my book, and so I was very enthusiastic. I thought: "I do not expect all one billion of the people in China to read my book, but even one-tenth of one per cent would do a lot for my readership". (Stiglitz)

The role is played on a two-plane system: as a rhetorical choice sometimes it is expressed in forms of low-profile and understatement, applying an argumentative reservation technique. Galbraith presents his "*small* book" as "an *attempt* to put the development task into clear and simple perspective" to "make some *small* contribution" to the understanding of the issues at stake. He says: "I do not suppose that *everything here will encounter the agreement* of all who read this book". Solow, on his part, says: "My own *inclination* – it is just an *inclination* – is to try a slightly different slant". And Stiglitz echoes: "*My views are offered up in the hope* that the world will be able to take this course of thinking through what the problems are and how we can fix them".

The words in italics, instead of reducing the speakers' role, produce the opposite effect, especially if put them in relation to the other sets of utterances which constitute the second, parallel plane of communication. Galbraith implicitly collocates himself in the group of the most important economists:

(15) Although one should always be wary of comparisons, this discussion can be compared in vigor with that which got under way after the publication of Smith's inquiry into "The Nature and Causes of the Wealth of Nations in 1776" and to which in the following sixty or seventy years Bentham, Malthus,

and John Stuart Mill, among others, made their notable contributions. The occasion is the same.

Solow describes himself as the economist able to change current trends, being an expression of his time:

(16) Keep in mind that Harrod's first Essay was published in 1939 and Domar's first article in 1946. Growth theory, like much else in macroeconomics, was a product of the depression of the 1930s and of the war that finally ended it. So was I. Nevertheless it seemed to me that the story told by these models felt wrong.

Stiglitz gives a full account of his professional achievements:

(17) For instance, when I was chief economist at the World Bank, I went to Geneva at the beginning of the next round of talks after the Uruguay Round and said: "Look, if you are going to have any hopes of succeeding in another round of trade discussions, you have to redress these imbalances".

Obviously, all this makes much more sense in relation to the scientific community, referred to as 'scholars', 'colleagues' and 'economists'. All of them animate the 'lived space' where they share linguistic forms, regulative norms and cultural concepts, and communicate content expertise according to discoursal expectations:

(18) It is our pride that the recent discussion of development has been scientific - that terms and concepts have been rigorously defined and so employed that <u>scholars</u> working on different aspects of the problem can communicate with each other with some certainty, can correct each other as necessary, and each by adding his piece of knowledge to the common stock can thus add to the total wisdom. (Galbraith)
If I may revert to methodological propaganda again, I would like to remind <u>my colleagues</u> and their readers that every piece of empirical economics rests on a substructure of background assumptions that are probably not quite true. (Solow)
The result of this has been that, in spite of the fact that <u>economists</u> know a lot more about how to manage an economy today than they did fifty years ago, there have been more than a hundred crises in the last three decades. (Stiglitz)

The relationship between the speaker and the scientific community is marked by intertextuality, which activates strategies of comparison, involving time and space:

(19) Both in the new states and in the old states it has been recognized that economic development is an imperative. Indeed, this has been a distinguishing feature of the recent as compared with the earlier discussion. (Galbraith)

When I look back now at the articles I wrote in the 1950s and 1960s on this general subject, I am struck and even a little surprised at how much effort went into broadening the technological framework of growth theory. (Solow)

The Fund actually tried to change its Charter in 1997, in the meeting held in Hong Kong. I had asked them at the time, when I was chief economist at the World Bank, where is the evidence that this change would improve the well-being of the developing countries? We knew that it would lead to more risk. But would it lead to more growth? (Stiglitz)

Although following sets of conventions given by the scientific community, the speaker expresses his personal authority in the forecast for the future in a continuous recontextualization of meaning with reference to time and space, since the present is embedded in the past and a prediction looks ahead with knowledge of the past (according to the Boroditsky's ego-moving schema, 2000: 5):

(20) To see the countries of the world not as divided between the developed and the underdeveloped but as spaced along a line representing various stages of development is essential for obtaining an accurate view of the problem of assistance. An equally complex organization would be a major misfortune for one of the newer African states with, for the foreseeable future, a far simpler range of tasks. (Galbraith)

But if one looks at substantial more-than-quarterly departures from equilibrium growth, as suggested for instance by the history of the large European economies since 1979, it is impossible to believe that the equilibrium growth path itself is unaffected by the short- to medium-run experience. (Solow)

The issue is not whether globalisation is going to change; globalisation will change. (Stiglitz)

The creation of new spaces relates directly to the topics of the lectures and covers different aspects; in Galbraith the introduction of new issues in the field of economics represents the new area of interest:

(21) Economic development in our day has come to be regarded overwhelmingly as a problem in economic analysis.

In Solow technology *is* the new space:

(22) A policy to increase investment would thus lead not only to higher capital intensity, which might not matter much, but also to a faster transfer of new technology into actual production.

In Stiglitz new spaces are occupied by new types of corporation:

(23) Now for the first time there are some new global monopolies, like Microsoft, which make entry into certain areas even more difficult.

The following examples show how each speaker uses his point of view, expressing his personal ideology, to engage also in phatic communication in order to establish a direct relationship with the audience, encouraged to know more about development, technology and global monopolies. Moreover, the speakers never forget they are part of the institution and are well aware of their personal role:

(24) It is our pride that the recent discussion of development has been scientific - that terms and concepts have been rigorously defined and so employed that scholars working on different aspects of the problem can communicate with each other with some certainty, can correct each other as necessary, and each by adding his piece of knowledge to the common stock can thus add to the total wisdom. (Galbraith)
In this brief review of the goals and achievements of growth theory I have referred as much to the work of others as to my own. That is more than mere modesty: the choice reflects my belief that any successful line of economic analysis is almost certain to be a group product. We attach names to ideas for good and bad reasons, but useful ideas are usually worked out and critically refined by a research community. (Solow)
The current system simply cannot continue. The issue is really whether it will change in the face of a crisis, or a series of crises, like the crises we had in 1997 and 1998, lurching from one patch-up to another, with each patch-up leading to a crisis maybe two years down the line, maybe ten years down the line, or whether we should approach it from the point of view of trying to understand why globalisation has not been working and how we might fix it. (Stiglitz)

6. Final remarks

We have seen that lectures call up models and frames, coordinate large arrays of information, and engage in creative mappings, transfers, and elaborations. There is an attempt in this contribution to show how the topics of the lectures considered (development, growth and globalization) imply an evolution in time and a positioning in terms of a place, an economic model or the scientific community.

The Boroditsky ego-moving schema and its sequence of past-present-future in lectures is functional to the construal of personal identity, to the history of the discipline and the socio-political events involved. The orientational metaphor underlying the speeches is the contrast in/out, in time or space. This metaphorical concept is not restricted to any physical experience (a particular time or a definite space) but belongs to the cultural knowledge of the speaker and the relevant audience which he is addressing, as in the following paragraph:

(25) The nineteenth century discussion was in a world that was rather proud of what was happening. The twentieth century discussion is in a world which feels that a great deal more must happen and very soon. (Galbraith)

The concept of inclusion/exclusion, a way of mapping between mental spaces, is also used with reference to the debate within the scientific community, giving the speaker the opportunity to reinforce his ideology, as his institutional status allows him to express a critical statement, as in the following:

(26) To leave it at that, however, to believe as many American economists do that empirical economics begins and ends with time series analysis, is to ignore a lot of valuable information that can not be put into so convenient a form. (Solow)

The contrast up/down or positive/negative, also present in the lectures, requires comparisons as connections between mental spaces: it is necessary, for example, to establish a comparison within the categories of time and space to understand the meaning of rich/poor (or, in other cases, of east/west). A form of evaluative language ("the lion's share", in the example which follows, does express a value judgment) is also necessary for the speaker to express his stance in the academic institution:

(27) For instance, take the Uruguay Round of trade discussions, which were completed in 1994. The result of that agreement was just as everybody had expected: the richest countries of the world (the United States and the EU) got the lion's share of the gains – but what they did not expect was that the poorest countries would actually be worse off. (Stiglitz)

Conceptual organization, founded on shared knowledge and aimed at construing identity, is the result of the speaker's role within his institution. Although an expression of the academic community, lectures delivered on special occasions by outstanding figures can overcome the borders of the scientific audience attending the event and reach specialists of other disciplines, students or anybody else. The speakers are aware of their role, as lectures are meant to transfer knowledge and take part in a process of popularization which is probably not expected of the genre. The discourse practices of academic institutions, however, come into play to guarantee a set of shared values, a basic consensus on what is to be discussed and how, and a common background of learning. Lecturers are able to embody time as the accumulation of institutional practices layered in cultural space; as Stiglitz says: "There have been very large changes both in the world and the world of ideas". Therefore, lectures can really perform a mediating function, establishing correspondences between time and space in multiple connections, in different contexts, exploiting the same cultural resources as an institutional wealth or, as in Solow's words, "the fund of common knowledge of the profession". And the economists have to take responsibility on their own shoulders: "in the *next world* I face quite a few years in that well-populated part of the *purgatory* where, it is known, economists are made to answer for all the advice they have given to governments" (Galbraith): so, a 'new' space will be provided for economists in a future time.

References

Benwell, Betham / Stokoe, Elizabeth 2010. *Discourse and Identity.* Edinburgh: Edinburgh University Press.

Bhatia, Vijay K. 2004. *Worlds of Written Discourse.* London: Continuum.

Bhatia, Vijay K. 2008. Towards critical genre analysis. In Bhatia, Vijay K. / Flowerdew, John / Jones, Rodney H. (eds) *Advances in discourse studies.* London: Routledge, 166-177.

Boroditsky, Lera 2000. Metaphoric structuring: understanding time through spatial metaphors. *Cognition*, 75, 1-28.

Bucholtz, Mary / Hall, Kira 2005. Identity and interaction: a sociocultural linguistic approach. *Discourse Studies* 7(4-5), 585-614.

Croft, William / Cruse, D. Alan 2010. *Cognitive linguistics*, Cambridge: Cambridge University Press.

Edwards, Derek 2006. Discourse, cognition and social practices: the rich surface of language and social interaction. *Discourse Studies* 8 (1), 41-49.

Fairclough, Norman 1992. *Discourse and Social Change*, Oxford: Polity Press.

Fairclough, Norman 2003. *Analysing Discourse: Textual Analysis for Social Research.* London: Routledge.

Fauconnier, Gilles 1997. *Mappings in thought and language*, Cambridge: Cambridge University Press.

Foucault, Michel 1969. *L'Archéologie du Savoir.* Paris: Gallimard.

Galbraith, John K. 1962. *Economic development in perspective*, Cambridge: Harvard University Press.

Galton, Antony 2011. Time flies but space does not: limits to the spatialisation of time. *Journal of Pragmatics* 43, 695-703.

Golshaie, Ramin 2011. Representing mental spaces and dynamics of natural language semantics. *BRAIN, Broad Research in Artificial Intelligence and Neuroscience* 2(1), 44-58.

Gotti, Maurizio 2012. Introduction. In Gotti, Maurizio (ed.) *Academic identity traits.* Bern: Peter Lang, 9-22.

Hyland, Ken 2006. Disciplinary differences: language variation in academic discourses. In Hyland, Ken / Bondi, Marina (eds) *Academic discourse across disciplines.* Bern: Peter Lang, 17-45.

Hyland, Ken / Bondi, Marina (eds) 2006. *Academic discourse across disciplines.* Bern: Peter Lang.

Lehrer, Adrienne 1990. Polysemy, conventionality, and the structure of the lexicon. *Cognitive Linguistics* 1, 207-246.

Salvi, Rita 2007. Listening to the writer's voice: the role of narrative interludes in argumentative discourse. In Garzone, Giuliana / Salvi, Rita (eds) *Linguistica, linguaggi specialistici, didattica delle lingue.* Roma: CISU, 215-225.

Salvi, Rita 2011. Reflections on genre and rhetorical choices in the language of economics. In Di Sabato, Bruna / Mazzotta, Patrizia (eds) *Linguistica e Didattica delle lingue e dell'inglese contemporaneo.* Lecce: Pensa Multimedia, 203-212.

Salvi, Rita 2012. Transfer of knowledge and academic discourse: lectures on economics. In Diani, Giuliana / Preite, Chiara (eds) *La trasmissione del sapere nelle diverse comunità accademiche: una prospettiva plurilingue.* Roma: Officina Edizioni, 75-100.

Salvi, Rita (forthcoming). Exploring political and banking language for institutional purposes. Proceedings of the CERLIS Conference 2012, University of Bergamo.

Soja, Edward 2001. *Postmodern geographies. The reassertion of space in critical social theory*, London: Verso.

Solow, Robert 1987. Growth theory and after. www.nobelprize.org

Stiglitz, Joseph E. 2008. Making globalization work. *The Economic and Social Review* 39 (3), 171-190.

Stubbs, Michael 2010. Three concepts of keywords. In Bondi, Marina / Scott, Mike (eds) *Keyness in Texts.* Amsterdam: John Benjamins, 21-42.

Tajfel, Henry / Turner, John 1986. The social identity theory of intergroup behaviour. In Worchel, Stephen / Austin, William G. (eds) *Psychology of Intergroup Relations.* Chicago: Nelson, 7-24.

Traugott, Elizabeth C. 1978. On the expression of spatio-temporal relations in language. In Greenberg, Joseph H. (ed.) *Universals*

of human language: Vol. 3. Word structure. Stanford, CA: Stanford University Press, 369-400.
van Dijk, Teun A. 2003. The discourse-knowledge interface. In Weiss, Gilbert / Wodak, Ruth (eds) *Critical discourse analysis. Theory and interdisciplinarity.* New York: Palgrave Macmillan, 85-109.

Giuliana Diani

The Construction of Academic Identity in English and Italian University Lectures

1. Introduction

It is generally acknowledged that academic lectures represent the principal genre of instruction, a crucial means to convey to students the knowledge base of a discipline (Flowerdew 1994; Thompson 1994; Young 1994; Flowerdew/Miller 1997). However, recent research shows that university lectures are more than a genre of knowledge dissemination. They are value-laden discourses in which lecturers not only transfer information to the audience, but they also express their stance towards the topic explained or discussed (Fortanet-Gómez 2004a; Biber 2006) by persuading the audience about the validity of their claims, and at the same time by contributing to the construction of a voice for authorial positioning in the scientific community. Lecturers' voice or *persona* (Cherry 1988) is constructed by means of different linguistic devices (e.g. the use of first person pronouns) which help lecturers project an impression of themselves and of how they stand in relation to their knowledge claims, their community, and their students.

A substantial amount of research carried out since the late 1990s has shown that personal pronouns play a crucial role in the construction of the textual identities of writers and readers (Ivanič 1998; Vassileva 1998; Bondi 1999; Kuo 1999; Tang/John 1999; Hyland 2001, 2002a, 2002b, 2012; Poppi 2004; Harwood 2005a, 2005b; Fløttum/Dahl/Kinn 2006; Lorés-Sanz 2006; Mur Dueñas 2007; Zareva 2013). Attention has also been paid to person markers in academic speech (Rounds 1987; Benwell/Stokoe 2002; Samson 2002; Fortanet-

Gómez 2004a, 2004b; Walsh 2004, 2010; Biber 2006; Crawford Camiciottoli 2007; Salvi 2012).

The aim of this paper is to contrastively analyse first person pronouns used by lecturers to project an authorial identity, which occur in the specific context of English and Italian lectures in the field of economics. I intend to explore whether their use is susceptible to cross-cultural variation.

My interest combines a focus on contrastive rhetoric (Mauranen 2001; Connor 2002, 2004) with an interest in linguistic resources in English and Italian. From this perspective, a comparative dimension of the analysis seems important, as it would allow for the study of linguistic variation across cultures.

In the following, a description of the comparable corpora analysed and the methodological procedures followed are outlined in section 2. Section 3 presents the results drawn from the analysis of the corpora under investigation. Finally, in section 4 some concluding remarks are offered in the light of the results obtained.

2. Materials and methods

The analysis is based on two small comparable corpora of university lectures in the discipline of economics and in different cultural contexts (English and Italian). The corpora include six lectures each. The English lecture corpus (totalling 54,583 words) consists of two lectures extracted from the *MICASE* corpus (*Michigan Corpus of Academic Spoken English*) (Simpson *et al.* 2002), and four have been collected at New York University, which were kindly given me by prof. Belinda Crawford Camiciottoli (University of Pisa, Italy). The lecturers were all native speakers of American English and the student audiences were also native speakers. The Italian data were assembled by the author at the University of Modena and Reggio Emilia (totalling 30,186 words).

The methodology adopted for this study combines a discourse and a corpus perspective. Discourse analysis contributes to the definition of pragmatic functions of person pronouns expressing the speaker's self (English *I* and 'authorial' *we*, *we* for *I*, Wales 1996), and their Italian counterparts, *io* and *noi*), whereas corpus linguistics offers ways of looking at lexical patterns: in particular, using *Wordsmith Tools* Version 5.0 (Scott 2008), I studied wordlists and concordances. These were worked out by comparing corpora to each other. Since the aim was to identify only those tokens of the lecturer's self-representation, a manual filtering process was necessary to make sure that only such occurrences were included. For example, instances of the English pronoun *we* and its Italian counterpart *noi* used to refer to both lecturers and student listeners were excluded, as they are considered to be inclusive engagement markers, thus performing a different function. The context was carefully analysed to include only those instances that refer to the lecturer(s).

English and Italian differ in the way first person pronouns are encoded in grammar. In English, subject pronouns are indispensable to identify the agent/actor of a given process. In Italian, personal pronouns may be used, but they are most often omitted, as the subject is always signalled through morphology in the verb-ending. For that reason, a computational search for the Italian verb-endings **amo* (present indicative)/**emo* (future indicative)/**emmo* (conditional) was carried out to obtain all tokens of verbs in the first person plural in the Italian corpus. But that was not done for the first person singular as it has different forms of verb-ending. A thorough reading of the corpus was thus necessary to find its occurrences. So, although in this particular realisation of person markers pronouns are compared with verb-endings, the comparison is still valid as they can be considered functionally equivalent.

The quantitative analysis was supplemented with qualitative analysis, looking at implications/discourse function in English and Italian.

3. Lecturer identity in English and Italian university lectures

3.1. Exploring frequency data

In this section, I sketch the main quantitative results of cross-linguistic comparison between *I/io* and *we/noi* used as instances of lecturer identity. Table 1 presents the data for first person pronouns in the two language corpora. Rather than indicating raw frequency counts, the data are reported according to the normalized parameter of tokens per 1,000 words. In this way, it is possible to make meaningful comparisons between corpora of different sizes (i.e. word counts).

First person pronouns	*EN lecture corpus (tokens per 1,000 words)*	*IT lecture corpus (tokens per 1,000 words)*
I / Io	3.05	2.18
authorial we (we for I) / noi	0.98	0.86

Table 1. Frequency of first person pronouns across the language corpora (1,000 words).

On the basis of this quantitative investigation, it can be said that the strategy preferred by Anglo-American lecturers is to use the first person singular pronoun *I* to construct their authorial persona (a total of 3.05 occurrences per 1,000 words). As regards the use of *we*, the data indicate that it is rarely used in the place of *I* (0.98 total occurrences).

Cross-cultural comparison shows that in Italian, too, the first person singular pronoun is the favoured choice to project an authorial identity (2.18 total occurrences vs. 0.86 occurrences of *noi* for *io*). However, this preference is less marked than in English. This could perhaps be explained in terms of the possible different conventional, traditional views on self-representation pertaining to the different national, big cultures (Atkinson 2004). Cultural traditions and conventions in the use of self-references are less strong in the Italian classroom teaching context. Thus, it may be argued that what contributes to establishing a credible ethos and, consequently, what constitutes an appropriate self-representation of the speaker may differ across different big or national cultures; these differences, in turn, may influence the frequency of use of self-mentions in spoken registers in general and in lectures in particular.

Quantitative data need to be complemented by a qualitative analysis targeting the pragmatic nature of this kind of manifestation. In the following section I explore the discourse function the lecturer takes on when referring to him- or herself by means of first person singular or plural pronouns.

3.2. Pragmatic nature of the lecturer's self-manifestation across the two language corpora

In order to investigate the discourse function of lecturer manifestation, I looked at the immediate co-text where the pronouns *I* and authorial *we* occur, i.e. the verb which is combined with them. My focus here is on cross-cultural variation which can be looked at both across languages and within the same language.

3.2.1. First person subject pronouns in English: a focus on co-text

Table 2 indicates the percentage distribution of verbs associated with *I* and authorial *we* in the English lecture corpus, with a frequency higher than 10.

I + *verbs* (freq. ≥ 10)	%	*authorial* we + *verbs* (freq. ≥ 10)	%
think	25.14	be gonna	42.59
wanna	10.17	can	31.48
be gonna	8.38	could	25.92
can	6.58		
know	5.98		
assume	3.59		
could	2.99		
expect	2.99		
want	3.99		
guess	2.99		

Table 2. Percentage distribution of verbs most frequently used in combination with *I* and authorial *we* in the English lecture corpus.

The figures show that economics lecturers rely heavily on verbs of stance. The range is related to the presence of stance verbs like *think, know, want, wanna, guess*, and also to the use of some modal and semi-modal verbs like *can, could, be gonna.* If we consider the verbs accompanying authorial *we*, choices are limited to modal and semi-modal verbs like *can, could, be gonna.*

Interestingly, the list of verbs combining with *I* also includes verbs like *assume* and *expect* that point to the central role of theoretical reasoning in economics discourse. When English economics lecturers use *I* in combination with these verbs, they highlight the importance of a 'rhetoric of inquiry' (McCloskey 1985; Klamer *et al.* 1988) that may be described as typical for economics.

Let us look at these verbs in more detail. Considering the verbs most frequently used with *I* in the corpus, *think* is the most frequent verb in percentage terms (25.14%). *Think* accounts for the greater number of epistemic stance markers in all lectures under investigation. Here is an example taken from the corpus (my bold as in the other extracts):

(1) But **I think** that uh the the the solution to it is in understanding the difference between what is what goes into the concept of the factor of production in economics. [...] (EN corpus)

It should be noted that not all occurrences of *think* have been considered. For example, *I'm thinking* was excluded, since it expressed the cognitive process of thinking, rather than 'giving an opinion'. However, although the most usual meaning of *think* in the corpus is 'to provide an opinion', sometimes its use seems to be nearer to a filler, with the function of giving the lecturer time to think about what she or he wants to say, as in the following excerpt:

(2) So that's why profit in the sense of the residual is often mixed together with management wages and with interest. It's easy to get these confused. Knight separated them out very carefully. Clark **I think** gets residual profit and interest and management wages all mixed up. (EN corpus)

Although this analysis could be applied to several other occurrences of *I think* in the corpus, it is difficult to distinguish their use from the one in which *I think* expresses an opinion. For this reason, all these occurrences have been included in the analysis.

Tokens of *I think* drawn from the computational analysis were closely examined to determine their discourse function. The analysis reveals that *I think* tokens were used by economics lecturers mainly to comment on the content of what is being said, thus conveying personal attitudes about course content, as illustrated in the following excerpts:

(3) The question is does ethics matter or do do particular ethical constraints matter for the marginal decision at hand. And that's where a lot of times it doesn't. And **I think** he fails to see that uh uh clear enough of the because of the as I say because of the uh the confusion as to what economists are trying to explain because models as you see you can solve for levels. (EN corpus)

(4) **I think** the recession is uh isn't productivity based it's based on gross national product isn't it? As long if the GNP doesn't grew for three consecutive months **I think** that's considered a recession. Well this isn't really GNP but uh but it does show that uh productivity is still there [...] (EN corpus)

Similarly, instances of *I know/I don't know* and *I guess* are used in the corpus as personal epistemic stance markers. For example,

(5) [...] Or they can pay it off early so you can see if if interest rates were going **I don't know** which way it would be up or down but then this is advantageous

if they got a buncha money they could go ahead and retire those and pay people off early still paying the thousand bucks they'll just pay it sooner [...] (EN corpus)

(6) Uh there's no special fees other than that that's that's the only one **I know** of no taxes to pay uh and course you're the owner you have complete control of what you do [...] (EN corpus)

(7) […] so **i guess** my claim is more, you know, how would you like to, crank a little more on the option value model and my my my sense is that, this is a dimension at least that you wanna think about. (EN corpus)

The verb *know* is above all frequent in the negative form. The frequent use of *I don't know* in the lectures under investigation is confirmed in related work on the use of multi-word sequences in classroom teaching. For example, *I don't know* is one of the most frequent sequences of words in Biber *et al.*'s (2004: 384) study of lecturers' use of lexical bundles in university teaching. Biber *et al.* (1999) discuss *I don't know* and combinations with *I don't know* as lexical bundles or "recurrent discourse building blocks, i.e. bundles of words that show a statistical tendency to co-occur" (1999: 991).

Attitudinal meanings are also expressed in the corpus through the combinations *I* + the lexical verb *want* and its phonological reduction *wanna.* These verbs are present in the corpus with the role of framing the lecturer's self-motivated desire in classroom teaching, as illustrated in the following excerpts:

(8) I'll take up a few uh additional points about Kirzner's theory of um uh alertness entrepreneurship and its relation to the issue of the justice of uh distributive shares and **I wanna** say a few things about the number of issues in uh the Sen book. (EN corpus)

(9) [...] So you see in in the latest of version of the minimum wage you read about in the textbook is that they that they realized this is gonna be the case so they they established a training wage. That is that if it's just ya know if it's your first job then the employer can pay you below the minimum wage at a certain level. So this this is to try to minimize the adverse impact on the teenagers. You know this this was recognized. So this I and see **I wanted** just uh to mention this that that um [...] (EN corpus)

However, a closer consideration of these occurrences shows that they are often used for functions other than the expression of personal desire. For example, the following excerpt illustrates the use of *I want* and *I wanna* for discourse organization, i.e. signalling to students what is about to happen in the lecture:

(10) [...] model, which uh, we started talking about last time so the idea is to uh, to bring in uh, the world economy after i finish this, uh **i wanna** start talking about what's called the Static Neoclassical Model. and, then next time **i wanna**, begin working on the second page of roman numeral one [...] uh on Thursday, **i want** to uh, talk about the Lucas Model, which has a paper and a little bit of reading in the textbook. and then uh next week on Tuesday **i wanna** spend one lecture on the Phillip's Curve, there's a handout on that there's a little bit of reading in the textbook, uh, that introduces some ideas about differential equations, and then uh next Thursday **i wanna** finish up, talking about the Dornbusch Model which is an elaboration on the Mundell-Flemming Model [...] (EN corpus)

Similarly, some modal verbs used in combination with *I* in the corpus like *will* and *be gonna* are used to identify the upcoming steps in the class session or some future action, as in:

(11) If we try to measure the quantity controlling for the fact that prices may change over time we call that idea real GDP. And **I'm gonna** do an example in a minute that I hope clarifies the difference between nominal and real for you. In order to measure GDP we need to have some idea about prices. (EN corpus)

(12) [...] that's what this is showing, same thing for capital... alright well so much for getting to the Static Neoclassical Model **i'll, i'll** do that next time, using this aggregate production function. uh, and **i'll** also talk about the Lucas Model [...] (EN corpus)

Returning to Table 2, another frequent verb accompanying *I* is the modal verb *can*. The occurrences of *I can* are used in the corpus mainly to indicate the lecturer's ability to carry out a procedure (and guide students through the steps of explanations), rather than expressing core stance meanings relating to epistemic possibility or personal permission. For example:

(13) […] if i started at time T, it's now time T plus delta. **i can** put a T on this one […] (EN corpus)

(14) okay now i'm gonna subtract this second expected value from the first because that's how i defined the option value function. and the reason that i broke this up into two pieces is so **i can** tr- subtract the, first piece of this second, from the first piece of the first equation […] (EN corpus)

The modal *could* was similarly used in the corpus, as in (15) and (16):

(15) let's think about this for a minute. If uh, E, changes smoothly, as a function of time, **i could** make a first-order approximation […] (EN corpus)

(16) […] and so if i knew the four key parameters **i could** calculate the probability that each individual would retire at age T […] (EN corpus)

The presence of instances of *can* and *could* in the corpus may be related to the discipline. The 'rhetoric of economics' can be shown to rely on mathematical reasoning (McCloskey 1985). The examples above show that the lecturers as economists focus on speculation and mathematical abstraction. Further examples from the corpus will be used to illustrate the high incidence of both speculative and hypothetical moves in the economics lectures under investigation.

As we will see in the following extracts, economics lecturers present hypothetical model-worlds with their data (*I'm assuming*, *I'm expecting*). They introduce their students to the use of mathematical models of economic behavior.

(17) **i'm assuming** that this, R, adjusts, and uh P and P-F are exogenous so **i'm assuming** that the E implicitly adjusts, to make that foreign exchange rate, uh, or to make the foreign exchange market, in my picture from the first day, balance. so that the inflow and outflow of dollars inflow and outflow of pounds from that, foreign exchange market, balance. yes... so there's the model. nine elements, three endogenous variables, three conditions here for equilibrium […] (EN corpus)

(18) so the comparison that i wanna make is between R, and R-F plus, E dot over E. and those are the two arguments here. and as was pointed out, we're talking about net, foreign financial investment in the U-S, when the interest rate in the U-S is is higher that's gonna be attractive, that's gonna pull money in, financ-

ing in, when uh R-F plus uh, E dot over E is high, that'll take it out. remember, the way I've defined this exchange rate, when this goes up, the dollar is devalued. if this is positive, it's it means **i'm expecting** the dollar to lose value. so if the, if the, interest rate in England is uh four percent, but **i'm expecting** the dollar to be devalued by six percent, i could have a ten percent return if i invested my money over there. cuz if the dollar is gonna lose value [...] (EN corpus)

Let us now consider the occurrences of authorial *we*. While the results offer a wider variety of verbs accompanying *I*, a very limited range of verbs accompany authorial *we* in the corpus, including only modal and semi-modal verbs. The semi-modal *be gonna* proves to be the most frequent choice (42.59%), followed by *can* (31.48%), and *could* (25.92%). For example:

(19) [...] Investment means something very specific here. It means purchases of capital equipment or capital structures that help you out in your business. And **we're gonna talk** about that more specifically later [...] (EN corpus)

(20) [...] and that's what **we're gonna** start on next week (EN corpus)

(21) [...] so the L-M curve shifts... so the equilibrium shifts like that, what happens, Y gets bigger, this i just, tried to say verbally. R, the exchange rate, the dollar devalues, and we have, expansionary monetary policy that happens, the interest rate doesn't change. the reason Y can get bigger, is because when the dollar devalues uh it stimulates exports. net exports. [...] so **we can** differentiate R with respect to M directly, R is just equal to a constant so, D-R is just zero. (EN corpus)

(22) you take the partial of this with respect to K it oughta be zero so **we could**, divide each side of that by P i could rewrite that then as this. (EN corpus)

One interesting aspect is that all the instances of authorial *we + be gonna* in the corpus are used for discourse organization rather than the more expected attitudinal meaning. This confirms results found in other studies of lecture discourse (Walsh 2004).

3.2.2. First person subject pronouns in Italian: a focus on co-text

Moving on to the Italian corpus of lectures, we find a picture of greater divergence in the range of verbs. An analysis of the verbs accompanying the first person singular pronoun *io* reveals that the relevant choices found were more limited. If the English corpus suggests a fair balance between choices of lexical verbs and of modals following *I*, the Italian corpus shows that references to lexical verbs following *io* are by far preferred: more than half of the occurrences of *io* are largely dominated by the verb of volition *volere* [want] (55%), the rest is divided between the verbs *sembrare* [seem] (25%) and *pensare* [think] (20%).

Most occurrences of *volere* serve a discourse organizing function, that is, they have the function of announcing the lecturer's intentionality and purpose, as in:

(23) Quello che **vorrei** fare oggi è una introduzione [...] (IT corpus)
[what i would like to do today is an introduction to...]

(24) Oggi **vorrei** discutere [...] (IT corpus)
[today i would like to talk about...]

The same function was noted for many instances of *want* and *wanna* in the English lecture corpus. However, differences were found in the verb tense. While Anglo-American lecturers favoured the use of the present indicative *I want* and *I wanna* for statements of purpose, Italian lecturers preferred the conditional form *vorrei* [i would like] as an oblique, hedged form.

When we analyse the pragmatic use of the other lexical verbs accompanying *io* in the corpus like *sembrare* and *pensare*, we find a similarity between these verbs and those found in the English corpus (*think*, *know*, *guess*) in their discourse function. Like the English verbs, the Italian *sembrare* and *pensare* tend to be associated with the discourse function of commenting on the content course on the part of the lecturer. Here are some examples from the Italian corpus:

(25) **Penso** che questo fenomeno rappresenti un ambito estremamente rilevante [...] (IT corpus)
[i think this phenomenon represents an extremely relevant area...]

(26) [...] quello che **mi sembra** di poter sottolineare è che [...] (IT corpus)
[... it seems to me that we can emphasize that...]

However, the quantitative data show that verbal reference to the making of comments is much more limited in Italian than it is in English. One reason may be that Italian lecturers are less inclined to emphasise their subjectivity when making claims and presenting their opinions. They tend to be considerably less visible than their Anglo-American colleagues.

If we consider the verbs accompanying the first person plural pronoun *noi*, we find a similar picture across the two language corpora. Like the English corpus, the Italian corpus offers a very limited range of verbs following the first person plural pronoun, which is only related to the use of some modal verbs like *dovere* and *potere*.

(27) La prima cosa su cui **dobbiamo** fare una breve valutazione è [...] (IT corpus)
[The first thing we must consider is...]

(28) [...] **noi possiamo** distinguere tra due approcci e rispetto a questi vi indicherò qual è il migliore [...] (IT corpus)
[we can distinguish between two approaches and I'll show you the best one...]

4. Concluding remarks

The findings of this study, while conditioned by the limited size of the corpora, offer evidence of cross-cultural variation in the use of first person pronouns, as devices used by lecturers to project an authorial identity. The results indicate that the use of first person singular pronoun is a common resource used by both Anglo-American and Italian lecturers in their university classroom. However, the frequency of first person singular pronouns has been found to be much higher in the

English corpus than in the Italian one (3.05 vs. 2.18). Interestingly, the frequency difference between the first person singular and plural pronoun *io/noi* in the Italian corpus is outstanding (*io* 2.18 vs. *noi* 0.86), as it is in the English corpus (*I* 3.05 vs. *we* 0.98).

The lower number of first person pronouns in the Italian corpus than in the English one could perhaps be explained in terms of the possible different conventional, traditional views on self-representation pertaining to the different national cultures, in which the lectures are delivered. Italian culture may run counter to the use of features which emphasise one's authorial persona, whereas Anglo-American culture could prompt the use of self-mentions to a greater extent.

Besides the greater frequency of use of first person pronouns in the English corpus, the contrastive analysis has yielded significant differences in the range of verbs accompanying the pronouns throughout the lectures in English and in Italian. The English corpus offers a much wider variety of verbs than the Italian one, which is related to the presence of some evaluative and affective verbs (*think*, *know*, *guess*, *want*, *wanna*), to the use of some modal and semi-modal verbs (*can*, *could*, *be gonna*), and to verbs signalling hypothetical reasoning (*assume*, *expect*). In the Italian corpus, on the other hand, the lexical choice of verbs is more limited (*volere, sembrare, pensare, dovere, potere*).

It should be pointed out, however, that similarity has been observed between the two corpora in the discourse functions of personal pronouns. In both English and Italian economics lectures, first person pronouns tended to be associated with verbs signalling lecturers' intentionality or purpose (English – *want*, *wanna*, *be gonna*; Italian – *volere*) and verbs expressing personal opinions (English – *think*, *know*, *guess*; Italian *sembrare*, *pensare*). This trend confirms that university lectures are more than a genre of knowledge dissemination. Lecturers not only disseminate knowledge to their students, but they also express their stance towards the topic explained.

The specificity and small size of the corpora analysed here calls for more analyses that confirm the results obtained. Nevertheless, it can be tentatively concluded that the differences in frequency and distribution of use of personal pronouns through which a lecturer is portrayed as an authorial self in a lecture could depend on the specific national culture. It would be interesting to investigate if this might also depend on the academic discipline.

References

Atkinson, Dwight 2004. Contrasting Rhetorics/Contrasting Cultures: Why Contrastive Rhetoric Needs a Better Conceptualization of Culture. *Journal of English for Academic Purposes* 3/4, 277-289.

Benwell, Bethan / Stokoe, Elizabeth 2002. Constructing Discussion Tasks in University Tutorials: Shifting Dynamics and Identities. *Discourse Studies* 4/4, 429-453.

Biber, Douglas 2006. *University Language: A Corpus-Based Study of Spoken and Written Registers.* Amsterdam: John Benjamins.

Biber, Douglas / Johansson, Stig / Leech, Geoffrey / Conrad, Susan / Finegan, Edward 1999. *Longman Grammar of Spoken and Written English.* London: Longman.

Biber, Douglas / Conrad, Susan / Cortes, Viviana 2004. 'If you look at...': Lexical Bundles in University Teaching and Textbooks. *Applied Linguistics* 25/3, 371-405.

Bondi, Marina 1999. *English across Genres: Language Variation in the Discourse of Economics.* Modena: Il Fiorino.

Cherry, Roger D. 1988. Ethos versus Persona: Self-Representations in Written Discourse. *Written Communication* 5/3, 251-276.

Connor, Ulla 2002. New Directions in Contrastive Rhetoric. *TESOL Quarterly* 36/4, 493-510.

Connor, Ulla 2004. Intercultural Rhetoric Research: Beyond Texts. *Journal of English for Academic Purposes* 3/4, 291-304.

Crawford Camiciottoli, Belinda 2007. *Language of Business Studies Lectures: A Corpus-Assisted Analysis.* Amsterdam: John Benjamins.

Fløttum, Kjersti / Dahl, Trine / Kinn, Torodd 2006. *Academic Voices*. Amsterdam: John Benjamins.

Flowerdew, John 1994. Research of Relevance to L2 Lecture Comprehension: An Overview. In Flowerdew, John (ed.) *Academic Listening: Research Perspectives*. Cambridge: Cambridge University Press, 7-29.

Flowerdew, John / Miller, Lindsay 1997. The Teaching of Academic Listening Comprehension and the Question of Authenticity. *English for Specific Purposes* 16/1, 27-46.

Fortanet-Gómez, Immaculada 2004a. Verbal Stance in Spoken Academic Discourse. In Del Lungo Camiciotti, Gabriella / Tognini Bonelli, Elena (eds) *Academic Discourse – New Insights into Evaluation*. Bern: Peter Lang, 99-119.

Fortanet-Gómez, Immaculada 2004b. The Use of 'We' in University Lectures: Reference and Function. *English for Specific Purposes* 23/1, 45-66.

Harwood, Nigel 2005a. 'Nowhere has anyone attempted ... In this article I am to do just that'. A Corpus-Based Study of Self-Promotional *I* and *We* in Academic Writing across Four Disciplines. *Journal of Pragmatics* 37/8, 1207-1231.

Harwood, Nigel 2005b. 'We do not seem to have a theory... The theory I present here attempts to fill this gap': Inclusive and Exclusive Pronouns in Academic Writing. *Applied Linguistics* 26/3, 343-375.

Hyland, Ken 2001. Humble Servants of the Discipline? Self-Mention in Research Articles. *English for Specific Purposes* 20/3, 207-226.

Hyland, Ken 2002a. Authority and Invisibility: Authorial Identity in Academic Writing. *Journal of Pragmatics* 34/8, 1091-1112.

Hyland, Ken 2002b. Options of Identity in Academic Writing. *ELT Journal* 56/4, 351-358.

Hyland, Ken 2012. *Disciplinary Identities: Individuality and Community in Academic Discourse*. Cambridge: Cambridge University Press.

Ivanič, Roz 1998. *Writing and Identity*. Amsterdam/Philadelphia: John Benjamins.

Klamer, Arjo / McCloskey, Donald N. / Solow, Robert M. (eds) 1988. *The Consequences of Economics Rhetoric*. Cambridge: Cambridge University Press.

Kuo, Chih-Hua 1999. The Use of Personal Pronouns: Role Relationships in Scientific Journal Articles. *English for Specific Purposes* 18/2, 121-138.

Lorés-Sanz, Rosa 2006. 'I will argue that': First Person Pronouns as Metadiscoursal Devices in Research Article Abstracts in English and Spanish. *ESP Across Cultures* 3, 23-40.

Mauranen, Anna 2001. Descriptions or Explanations? Some Methodological Issues in Contrastive Rhetoric. In Hewings, Martin (ed.) *Academic Writing in Context*. Birmingham: University of Birmingham, 43-54.

McCloskey, Donald N. 1985. *The Rhetoric of Economics*. Madison, Winsconsin: University of Wisconsin Press.

Mur Dueñas, Pilar 2007. 'I/we focus on...': A Cross-Cultural Analysis of Self-Mentions in Business Management Research Articles. *Journal of English for Academic Purposes* 6/2, 143-162.

Poppi, Franca 2004. Pragmatic Implications of the Use of 'We' as a Receiver-Including and Receiver-Excluding Pronoun. In Aijmer, Karin (ed.) *Understanding and Misunderstanding in Dialogue*. Tübingen: Niemeyer, 229-242.

Rounds, Patricia L. 1987. Multifunctional Personal Pronoun Use in An Educational Setting. *English for Specific Purposes* 6/1, 13-29.

Salvi, Rita 2012. Transfer of Knowledge and Academic Discourse: Lectures on Economics. In Diani, Giuliana / Preite, Chiara (eds) *La Trasmissione del Sapere nelle Diverse Comunità Accademiche: Una Prospettiva Plurilingue*. Roma: Officina Edizioni, 77-102.

Samson, Christina 2002. Negotiating Academic Knowledge: The Use of Person Markers in Contemporary Written Economics Lectures. *Quaderni del Dipartimento di Linguistica* – Università di Firenze 12.

Scott, Mike 2008. *Wordsmith Tools*. Oxford: Oxford University Press.

Simpson, Rita / Briggs, Sarah / Ovens, Janine / Swales, John 2002. *The Michigan Corpus of Academic Spoken English*. Ann Arbor,

MI: The Regents of the University of Michigan. <http://www.lsa.umich.edu/eli/micase/micase.htm>.

Tang, Ramona / John, Suganthi 1999. The 'I' in Identity: Exploring Writer Identity in Student Academic Writing through the First Person Pronoun. *English for Specific Purposes* 18/S1, S23-S39.

Thompson, Susan 1994. Frameworks and Contexts: A Genre-Based Approach to Analysing Lecture Introductions. *English for Specific Purposes* 13/2, 171-186.

Vassileva, Irena 1998. Who Am I/Who Are We in Academic Writing? *International Journal of Applied Linguistics* 8/2, 163-190.

Wales, Katie 1996. *Personal Pronouns in Present-Day English*. Cambridge: Cambridge University Press.

Walsh, Polly 2004. A Complex Interplay of Voices: First and Second Person Pronouns in University Lectures. In Anderson, Laurie / Bamford, Julia (eds) *Evaluation in Oral and Written Academic Discourse*. Roma: Officina Edizioni, 31-52.

Walsh, Polly 2010. Multiple Roles and Footing Shifts in Academic Lectures. In Garzone, Giuliana / Archibald, James (eds) *Discourse, Identities and Roles in Specialized Communication*. Bern: Peter Lang, 383-405.

Young, Lynne 1994. University Lectures: Macro-Structure and Micro-Features. In Flowerdew, John (ed.) *Academic Listening: Research Perspectives*. Cambridge: Cambridge University Press, 159-176.

Zareva, Alla 2013. Self-Mention and the Projection of Multiple Identity Roles in TESOL Graduate Student Presentations: The Influence of the Written Academic Genres. *English for Specific Purposes* 32/2, 72-83.

GIUDITTA CALIENDO

Ago ergo sum. New Ideological Perspectives in the Discursive Construal of Institutional Identity

1. Introduction

This chapter presents the results of a longitudinal investigation on the role played by discourse in the representation of the European Union as a political and social actor. The study gives emphasis to the way institutional actions to pursue legitimacy and consensus take the shape of communicative strategies which directly affect citizens' perception of the EU as a political entity.

In its early days, the European Economic Community (EEC) was an organization constituted on purely economic grounds which aimed at setting up a common market and a customs union. As the EEC gradually evolved into what is today the European Union, Community institutions became increasingly interested in engaging the citizen in the activities of the EU. Still today, in an attempt to gain recognition and legitimacy, the European Union is eager to present its construction as a process of common and social growth, rather than a mere economic convergence. As Streeck (1995: 409) puts it, the creation of a 'Europe for the citizens' is paramount to ensure that the EU project will not die from lack of popular support. However, in the desired transition from the 'Europe of the markets' to the 'Europe of the people', the Union is facing one the hardest challenges in its history: becoming real to the citizens and making the EU project a participatory experience. In spite of all the EU's efforts to legitimate its governance and connect better to its citizens, opposition remains strong in numerous quarters and citizens are still sceptical of its political rationale (Calhoun 2003; Keohane/Nye 2003). The causes of discontent are numerous and seem to be associated to the EU's atypical structure (Citrin/Sides 2004; Schmitter 1996), its expanding and unstable borders (Taylor 2008), the lack of a common ethnic identity (Etzioni 2007) and its alleged technocratic nature (Goldsmith 2003).

In this respect, political scientist E.O. Eriksen (2009: 22-23) makes a highly interesting claim when he remarks that, in order to be conceived as a participatory and 'collective enterprise', the EU should explicitly offer citizens a justification of its mission, and become salient to their personal lives. In doing so, he envisages three possible functional models around which the EU positions itself and constructs its political identity: 'the EU as a problem-solver', 'the EU as a provider of common values', 'the EU as a provider of rights'. In this chapter, I argue that the emergence of these new identity roles is related to processes of discursive change which can be clearly traced in EU texts addressing the general public.

The corpus is based on a comprehensive collection of informative and popularizing booklets which the EU has published in the EUROPA website over a time span of approximately ten years. These publications were collected and classified in order to gather a fully representative sample of EU informative discourse addressing the general public. The criteria followed for the corpus selection are meant to highlight the way in which rhetorical and discursive strategies have undergone significant forms of change over the last decade, especially as a result of the 2005 referendum debacle in France and the Netherlands on the Treaty establishing a Constitution for Europe. The corpus analysis brings to light some major and interesting modifications in terms of discursive patterns, contents, as well as identity values being foregrounded and prioritised. The results are all the more interesting as these changes are perfectly in line with the objectives of a wide series of official action plans undertaken after 2005 and calling for a drastic reform of EU communication and information strategies in order to counter the democratic deficit.

2. Historical background

The referendum on the Treaty establishing a Constitution for Europe held in 2005 is considered by many as the moment when EU popularity and consensus were most under pressure. Although high hopes and grand visions were in the air at that time, the ratification process came to a standstill when the French and Dutch electorate expressed their dissatisfaction towards the EU by rejecting by a large majority the Treaty during the referenda called, respectively, on May 29 and June 1 2005. Since then, it became painfully clear to the EU that the need to strengthen its legitimacy and public endorsement was growing fast and could no longer be overlooked.

Soon after the referendum debacle a 'period of reflection' on the future of Europe was announced through a declaration of the European Council on June 18 2005. This period of reflection was meant to give EU countries more time to debate the Constitution and, more generally, to reconnect the citizens with the European project. A few months later, in October 2005, the European Commission presented a 'Plan D', where D stood for 'Democracy, Dialogue and Debate'. Soon enough, it became clear that 'D for Debate' implied an active involvement on the part of the citizens and that efforts to stimulate their reaction and participation had to be complemented by a better dissemination of EU-related information through a wide range of channels, including traditional media and new communication technologies. This gave new impetus to the debate on the future of Europe by encouraging new ways to draw citizens into the dialogue[1]. The idea was that making citizens more aware of the EU would help remove what was a 'communication deficit' rather than a 'democratic deficit'. This translated into attempts to provide accessible, reader-friendly documentation and more transparent information, which implied inevitable changes in the EU communication policy.

Against this background, many initiatives in the form of action plans were undertaken in the field of communication to counter public distrust and contribute to this overall period of reflection. As Table 1 below shows, a number of official documents were issued and initiatives undertaken with a view to: listening to the public and taking their views and concerns into account; explaining how institutional activities and EU policies affected their everyday lives; connecting with people locally and addressing them in their national or local settings, through their favorite media[2].

1 <http://www.mva.si/fileadmin/user_upload/doc/4_MLADINSKA_POLITIKA/3_Strukturirani_dialog/Plan_D.pdf> (Last accessed: October 2012)

2 <http://ec.europa.eu/dgs/communication/about/what_we_do/index_en.htm> (Last accessed: October 2012)

YEAR	COMMUNICATION INITIATIVE	AIM
2005	Communication to the Commission. Community Action Plan to improve communicating Europe by the Commission	50 concrete measures to change the Commission's communication approach
2005	Communication from the Commission to the Council, the European Parliament, the European Economic and Social committee and the Committee of the Regions – The Commission's contribution to the period of reflection and beyond – Plan-D for Democracy, Dialogue and Debate	Discussing the EU and its policies and the future of Europe with the broad public, interest groups, employers and trade unions, national parliaments and political parties
2006	White Paper on a European Communication Policy	Bridging the gap between Europe and its citizens
2007	Communication to the Commission. Communicating about Europe via the Internet to engage the citizens	Engaging the citizens via the use of new technologies
2008	Debate Europe[3]	Online forum where people can voice their concerns to decision-makers
2008	Communication to the Commission. Communicating about Europe via the Internet. Engaging the citizens	Making the EUROPA website the one-stop site for all EU institutions and information
2008	Communicating Europe through audiovisual media	Boosting coverage of EU affairs on new and existing audiovisual platforms
2007	Communication from the Commission to the European Parliament, the Council, the European Economic and Social Committee and the Committee of the Regions. Communicating Europe in Partnership	Closing the communication gap between the EU and its citizens through efficient cooperation and partnerships (follow-up of the 2006 communication document)
2009	Evaluation of the Plan D / Debate Europe citizen consultation projects. Final Report	Evaluation of the Plan D / Debate Europe citizen consultation projects.

Table 1. Communication initiatives undertaken by the EU after 2005.

The wide number of initiatives undertaken in the field of communication listed in Table 1 above shows that communicating with the public

3 <http://ec.europa.eu/archives/debateeurope/index_en.htm> (Last accessed: October 2012)

– always one of the EU institutions' key activities – became a primary concern for the European Union in the wake of the 2005 referendum failure. One document in particular, the *White Paper on a European Communication Policy* issued by the Commission in February 2006, even stated that communication should become an 'EU policy in its own right', at the service of all EU citizens. All these actions were moving in the direction of establishing a European 'public sphere', i.e. a space of exchange and relationships among citizens from all walks of life which "allows for the production, reproduction, or transformation of a social imaginary that gives cultural form to integration, making Europe real and giving it shape by imagining it in specific ways" (Calhoun 2003: 244).

3. Corpus construction and aims

For the purpose of this analysis, all the EU informative booklets published from 2000 to 2010 in the main informative section of EUROPA ('Easy Reading Corner') were collected. The collection can be considered comprehensive as these booklets, initially conceived in paper version, started to be circulated in their electronic format on the official EUROPA website in 2000.

Since this study is meant to highlight the way in which discursive and rhetorical strategies have undergone significant forms of change in the wake of the 2005 referendum, the corpus was divided into two subcorpora, as shown in detail in Tables 2 and 3 below:
- Corpus A: 27 booklets published prior to the 2005 referendum (time span: January 2000-June 2005);
- Corpus B: 37 booklets published after the 2005 referendum (time span: July 2005-December 2010).

Title and publication date	Tokens
ABC of Community Law (2000)	41,659
European employment and social policy: a policy for people (2000)	10,582
How Europeans see themselves (2000)	13,785
A new idea for Europe (2000)	15,244
Tax policy in the European Union (2000)	10,043
A community of cultures. The European Union and the arts (2001)	7,585
European solidarity with the victims of humanitarian crises (2001)	5,629
It's a better life* (2002)	6,255
Making globalisation work for everyone (2002)	5,873
Serving the people of Europe. How the European Commission works* (2002)	6,493
Towards a knowledge based Europe (2002)	5,488
Choices for a greener future* (2002)	9,929
Travelling in Europe* (2002)	5,315
Europe at a crossroads. The need for sustainable transport (2003)	5,554
Europe in 12 lessons* (2003)	23,182
Going for growth. The Economy of the EU (2003)	5,605
More unity and more diversity. The European Union's biggest enlargement (2003)	5,653
Panorama of the European Union* (2003)	3,123
How the European Union works. A citizens' guide to the EU institutions* (2003)	14,413
It's your Europe (2003)	5,379
Freedom, security and justice for all (2004)	5,580
Looking beyond tomorrow. Scientific research in the European Union (2004)	5,302
Many tongues, one family* (2004)	5,356
Working for the regions (2004)	13,127
From farm to fork (2004)	5,221
Key facts and figures 2005*(2005)	8,864
Panorama of the European Union* (2005)	3,147
TOTAL	253,386

Table 2. Corpus A: EU informative publications issued from January 2000 to June 2005[4].

4 Booklets marked by an asterisk in Tables 2 and 3 were edited and re-published after June 2005. For a comprehensive list, see the Appendix.

Title and publication date	Tokens
Serving the people of Europe. What the Commission does for you* (2005)	7,185
How the European Union works. Your guide to the EU institutions (2005)	15,181
Better off in Europe* (2005)	6,752
A quality environment* (2005)	5,341
Europe and you in 2006: a snapshot of EU achievements (2006)	2,561
Europe in 12 lessons 2006* (2006)	16,991
In touch with the EU. Ask your questions – have your say (2006)	3,673
Investing in our common future – The budget of the European Union (2006)	5,293
Travelling in Europe – 2006* (2006)	4,476
Combating climate change. The EU leads the way 2007 (2007)	5,474
The EU in the world. The foreign policy of the European Union (2007)	5,229
Europe and you in 2007: a snapshot of EU achievements (2007)	2,660
How the EU works. Your guide to the EU institutions* (2007)	12,954
Key facts and figures about Europe and the Europeans* (2007)	9,951
Panorama of the EU 2007* (2007)	2,845
Travelling in Europe – 2007* (2007)	4,896
50 ways forward. Europe's best successes (2007)	18,261
Your rights as a consumer. How the EU protects your interests (2007)	5,837
Combating climate change. The EU leads the way 2008 (2008)	5,532
Europe and you in 2008: a snapshot of EU achievements (2008)	2,384
Jobs and growth in the EU. A roadmap to a sustainable tomorrow (2008)	5,685
Speaking for Europe. Languages in the European Union* (2008)	6,254
Travelling in Europe – 2008* (2008)	4,403
EU information and assistance services (2009)	5,528
Europe for women 2009 (2009)	1,916
An opportunity and a challenge. Migration in the European Union (2009)	5,356
Panorama of the EU 2009* (2009)	2,822

The state of the Union in 2009: the key achievements of the Barroso Commission (2009)	8,528
Travelling in Europe – 2010* (2009)	4,180
Your guide to the Lisbon Treaty (2009)	4,564
Europe and you: a snapshot of EU achievements. (2010)	2,398
Europe for women 2010 (2010)	1,924
Europe in 12 lessons 2010* (2010)	19,443
The European Commission 2010-14. Profiles and priorities (2010)	9,537
Travelling in Europe – 2011-12* (2010)	4,226
Your guide to the Lisbon Treaty (2010)	4,564
Europe and you: a snapshot of EU achievements (2010)	2,398
TOTAL	237,202

Table 3. Corpus B: EU informative publications issued from July 2005 to December 2010.

In this study, based on quantitative and qualitative methods, EU informative materials are scrutinized in order to investigate the longitudinal evolution of discursive practices of self-representation employed by the EU over the time span 2000-2010. These strategies serve a number of purposes (e.g. representing social actors, strengthening roles and claims, conveying agency and accountability) and exploit key grammar devices such as agentivity, topicalization, deictic elements, reference and pronouns.

In order to understand the evolution of discursive conventions over time, the informative publications made available by the EU to the general public are also investigated in order to document instantiations of interdiscursivity (Fairclough 1992) and the emergence of traits of promotional discourse. As the analysis of more recent publications shows, EU discourse tends to be increasingly centred on the benefits, opportunities and practical results that EU membership brings about as a way to enter into direct contact with the citizens and become increasingly relevant to their daily life (Risse 2004).

The two subcorpora together represent the reference corpus, against which both subcorpora A and B have been searched. The collected data have been converted and edited to be analyzed with the language software WordSmith Tools 5.0 (Scott 2008) in order to allow identification of recurring linguistic structures by means of a statistical quantitative approach.

Although the investigation is based on a small corpus, the data emerging from the longitudinal analysis provide a clear-cut observation of some micro changes in EU discourse which prove particularly meaningful as they reflect key social and political conjunctures (Salvi forthcoming). It is also worth pointing out that about 40% of the booklets contained in corpus A have been edited and re-published after June 2005 (see Appendix). This aspect strengthens the contrastive nature of the analysis as the comparison of the two versions of the same booklet in corpora A and B enables to track down new discursive trends in a more direct fashion.

4. Theoretical framework and findings

This section is devoted to the analysis of the EU's discursive self-representation and focuses on the way institutional action plans in the field of EU communication policy translate into new rhetorical practices aimed to pursue consensus. The analysis of the informative booklets published by the EU over the decade 2000-2010 reveals the crucial role played by discourse in countering the lack of public support (Streeck 1995: 409) and creating a 'Europe for the citizens'. From a longitudinal perspective, the most recent publications of the EU seem to take on a more persuasive slant as the discursive strategies employed foreground the numerous advantages and benefits that EU membership brings about.

In this respect, Eriksen (2009: 22-23) rightly remarks that to be conceived as a 'collective enterprise', and not simply as a free market, the European Union should explicitly offer citizens a justification of its mission or vision, and legitimize its political entity in effective and tangible ways. In doing so, he provides a typology of the ways in which legitimacy may come about and envisages three possible strategies through which the EU can create its identity, justify its actions and, at the same time, repair its alleged legitimacy deficit. These three possible strategies of legitimation are based on utility, common values and rights, as is summarized in Table 4 below:

Type of entity	Conception of rationality	Mode of legitimation	Type of citizenship
Problem-solving entity	Instrumental (Legitimacy depends on the ability to solve problems effectively and to offer discernible benefits)	Efficiency	Economic
Value-based community	Contextual (Legitimacy depends on the ability to create a framework of shared common values)	Collective self-understanding	Social and cultural
Rights-based union	Communicative (Legitimacy depends on the ability to empower citizens and confer upon them a set of rights)	Justice and norms of fairness	Political

Table 4. Three logics of polity formation (adapted from Eriksen 2009: 61).

Each of these three different conceptions of legitimation is based on a specific logic which is, respectively, the instrumental, contextual and communicative one. According to instrumental logic, the EU is conceived as a polity whose main aim is to solve problems that Members States and citizens are challenged with and to offer tangible benefits associated with EU membership. In line with contextual logic, the European Union identity is based on collective self-understanding and on a series of identity-forming values. As far as communicative logic is concerned, this sees the EU as a union based on rights and ascribed to a specific legal framework (Eriksen 2009: 59-74).

Eriksen's three strategies of legitimation are suggested paths of institutional development to 'repair' legitimacy deficit and are discussed by the author as ideal types of polity formation. Since the main purpose of this paper is to tease out and explore shifts in discourse in response to a legitimacy crisis, these strategies will be analysed discursively so as to find out which aims are currently being pursued by the European Union in its communication policy and which mode/s of legitimation is/are being prioritized in discourse to increase consensus and public endorsement.

4.1. The EU as a problem-solving entity

According to this first suggested type of polity formation, the European Union's legitimacy is subject to the institutions' ability to solve problems effectively and to deliver the goods and services that people demand (Eriksen 2009). EU identity is in this case conceived from a functional and utilitarian perspective as citizens' allegiance is linked to the discernible benefits that derive from membership. This type of policy formation corresponds to the 'socialization model' of Hermann and Brewer (2004: 13-15) who illustrate linkages between institutions and their changing identities. They believe that an evolution in institutional identity affects the affiliation felt for that institution and that, as far as the socialization model is concerned, citizens come to identify increasingly with an institution to the extent that it is a trustworthy and effective entity, 'salient to their personal lives'.

Moving from a theoretical to a more experiential perspective, studies have demonstrated that, in practice, individuals do show greater endorsement for the EU when it comes to solving problems and that utilitarian motives strongly boost legitimacy and citizens' acceptance of the EU. An interesting study by Caporaso and Kim (2008) has shown that, even if citizens' identification with Europe remains weak, their level of support is astounding when asked questions concerning decision-making and problem-solving capacity in EU policy domain. In their study, based on the interpretation of Eurobarometers' data, a set of indicators are taken into consideration in order to assess changes in the way the EU is perceived and to show that citizens, by large numbers, support increasing delegation of functions to the EU. In particular, when analyzing interests and preferences, Caporaso and Kim demonstrate that there is a tendency on the part of the citizens to turn to the EU to solve a broad range of problems which were previously within the purview of the member states.

Against this background, and moving on to the analysis of the informative texts under investigation, it is expedient to underline that in corpus B, i.e. after the 2005 referendum, the European Union is described for the very first time as a 'problem-solver':

(1) The influence of *the EU as 'problem solver'* on the global stage has been brought to the fore in global efforts to tackle the economic crisis. (The state of the Union in 2009: the key achievements of the Barroso Commission. 2009)[5]

The analysis of the keywords of corpus B also illustrates that *the EU* emerges as the top keyword. The institutional actor therefore plays an active and dominant role in the narrative of the new set of informative booklets.

A more detailed look at the results emerging form the keyword analysis suggests that the higher frequency of *the EU* is also accompanied by its agentivization, which translates into a shift from object to subject position. This is all the more evident from the comparative analysis of the same booklet in its pre- and post-referendum version. Some interesting instantiations of this longitudinal change can be observed in Table 5, where examples (2) and (4) are compared to their post-referendum version (3) and (5):

Corpus A	Corpus B
(2) *The target for 2010 is* to generate 21% of the EU's electricity from renewable sources. (Key facts and figures about Europe and the Europeans. May 2005)	(3) *The EU has set a target* of generating 21% of its electricity from renewable sources such as wind, solar, hydro, geothermal and biomass by 2010. (Key facts and figures about Europe and the Europeans. 2007)
(4) *It is a high priority for the EU to help people move around* for work and pleasure, to speak to each other across borders and to reinforce their sense of belonging to a single Community. (Many tongues, one family. Languages in the European Union. 2004)	(5) *The EU is committed to supporting the rights of its citizens* to personal and professional mobility and their ability to communicate with each other. (Speaking for Europe. Languages in the European Union. 2008)

Table 5. Corpus A *vs.* corpus B with reference to cases of agentivization of *the EU*.

Of course agency means control and responsibility over actions being undertaken and the 'longitudinal shift' in agency shown in the examples listed in Table 5 demonstrates that the EU is increasingly portrayed as an active participant, a 'doer' who has the 'ability' to undertake measures and solve problems. What is central in the corpus of booklets published after 2005 is indeed the EU's 'ability' to 'take ac-

5 In all examples, emphasis is added.

tion', 'tackle challenges', 'spread European values' or 'fight international cross-border crime', etc. Interestingly enough, a comparison between the two corpora displays an interesting increase in the overall number of occurrences of the term *ability*:

	Corpus A	Corpus B
Ability	18	*32*

Table 6. Occurrence of the noun *ability* in the two subcorpora.

What is truly remarkable about the occurrence of the term *ability* is not only its numerical increase after June 2005, but also a shift in terms of the subject who possesses the ability. A more detailed analysis of the quantitative data shows that in the booklets published before June 2005 the noun *ability* was used with the EU as a subject or agent only in 16% of the cases. In the remaining occurrences, the term *ability* occurred with subjects other than the EU, mainly candidate countries for accession, and referred to their capacity to meet the obligations of EU membership:

(6) In December 1997, the process of enlarging the European Union to the south and the east was launched. Some 13 new countries have applied to join the EU. They should be in a position to join the Union as soon as they have demonstrated *their ability to assume the obligations of membership*. (How Europeans see themselves. 2000)

(7) Since its creation, Phare has concentrated increasingly on projects to help ensure that *the newcomers have the administrative ability to meet the rights and obligations of membership*. These projects now account for about 30% of the Phare budget. The idea is, first, to make sure that *they have the ability to make good use of the financial assistance the EU provides* to boost growth and jobs, and, second, to enable them to apply EU rules fully and correctly. (More unity and more diversity. 2003)

In the new corpus the scenario has completely changed: *ability* is mainly used with reference to the agent *the EU/European Union* (44% of the occurrences), presented either in the genitive case (example 8) or in '*of*-construction' prepositional phrases (example 9):

(8) This change will improve *the EU's ability* to act, especially in external affairs. The Lisbon Treaty will allow the EU to act more effectively, coherently and

credibly in its relations with the rest of the world. (Your guide to the Lisbon Treaty. 2010)

(9) But it also put the spotlight on what had made possible 50 years of achievement: *the ability of the EU* to develop through these changes on the basis of a common direction and common values — and as a result, to reach goals more successfully together than would be possible apart. (The state of the Union in 2009: the key achievements of the Barroso Commission. 2009)

As the analysis shows, the EU is increasingly constructing its narrative and discursive self-representation by putting forward its agency and its image as problem solver. In this way, the EU is discursively promoting the positive and beneficial results deriving from its actions so as to appear increasingly relevant to citizens' daily life (Risse 2004) and achieve proximity with them.

In the 'post referendum' corpus this 'discursive vicinity' to the citizens is also achieved by means of a remarkable increase in the use of agentive adjuncts expressing instrumentality, such as the complex preposition *thanks to*:

	Corpus A	Corpus B
Thanks to	43	*68*
Thanks to the EU	1	*12*
Thanks to the European Union	1	*10*
Thanks to + the EU- (as premodifier)	1	*8*
Thanks to + Europe's (as possessive agent)	1	-

Table 7. Occurrence and collocations of *thanks to* in both subcorpora.

As can be inferred from Table 7, in the corpus containing the new booklets the main collocate of *thanks to* is *the EU/European Union*:

(10) *Thanks to the European Union*, Europe now stands at the leading edge of the renewable energy trend, promising sustainable health, wealth and prosperity. (50 ways forward. Europe's best successes. 2007)

(11) *Thanks to the EU*, air travellers now have a series of rights (which must be displayed at EU airports) setting out airlines' obligations to customers when flights are delayed or cancelled and fixing compensation levels if flights are overbooked. (Better off in Europe. How the EU's single market benefits you. November 2005)

From a semantic perspective, *thanks to* does not only express agentivity but also evaluation, as it conveys a positive aura to the EU's different types of intervention which are emphasized and applauded.

4.2. The EU as a value-based community

According to the second strategy theorized by Eriksen (2009), the EU legitimizes its existence and entity through a set of shared common values which are meant to represent who the people of Europe are and who they would like to be. However, as Eriksen (2009: 69) suggests, it is far from clear what values and virtues are uniquely European, as opposed to universally shared or shared among smaller communities. Regardless of the context of reference, values are always an extremely complex issue and this is particularly true in the case of the EU, a space where so many different and diverse people converge.

The contrastive analysis between corpus A and corpus B interestingly shows a variation in terms of the premodifiers generally used to refer to the generic noun *value**. A first difference worth pointing out is the increase in the occurrence of the collocation '*common + value**' which more than doubles in corpus B. This increase can also be detected when observing the occurrence of similar types of nouns, such as *interest** and *goal**, which collocate with the pre-modifier *common* twice as much as in corpus B. The emphasis is therefore not so much on the types of values being presented (which are numerous and varied in both subcorpora, ranging from democracy to prosperity) as to their feature of being 'shared' and 'common to us all'.

Another interesting aspect emerging from the comparison between the two subcorpora in relation to the string 'premodifier + *value**' is also the disappearance of the genitive premodifier *citizens'* in the new booklets. The collocation *citizens' values* totally disappears in corpus B and is replaced by the collocation *Europe's values*, suggesting an idea of 'inclusiveness' and of pan-Europeanness that goes beyond the citizen, the individual.

In the new EU booklets contained in corpus B the idea of sharing a common set of values and a common view of the world is also emphasised by means of elements of personal deixis (Table 8) such as: first person plural form *we*; subject complement *us*; and the possessive form *our*. It is through personal deixis that the EU intends to portray itself as closer to the citizens, discursively creating a 'we-group' and a consequent sense of proximity and togetherness (see example 12).

	Pre 2005	Post 2005
We	176	*198*
Our	144	*190*
Us	24	*53*

Table 8. Occurrences of elements of personal deixis in the two corpora.

(12) The crisis has provided *us* with an opportunity to draw important lessons and decide how *we* want to build *our* economic landscape from here on; how best to support European businesses and enable them to prosper, while keeping social issues high on the agenda. (Europe for women. 2009)

If we look in more detail at the main clusters of the pronoun *we* (Table 9 below, trigram query), it emerges that its collocates change completely after 2005:

Corpus A	Corpus B
we look at	*we need to*
we are not	*we want to*

Table 9. Main clusters of *we*.

The occurrences of *we* in corpus A mainly express the highly referential nature of EU booklets (example 13) and play a metadiscursive function, leading the target readers through the text. To a less extent, *we* is used in concessive clauses, like the one exemplified in (14):

(13) In this booklet, *we look at the wide array of languages spoken in the EU*, at the choice and challenge of learning a second (or third) language, at the EU's programmes for language teaching and language learning, and finally at what it takes to run a multilingual Union. (Many tongues, one family. Languages in the European Union. 2004)

(14) *Although we are not* always aware of it, each one of us in day-to-day life enjoys the benefits of the single market: access to a wide range of consumer goods and products; prices kept down by competition; policies that protect

consumers and the environment; technical standards that tend to be harmonised upwards. (Europe in 12 lessons. 2003)

By contrast, when observing the clusters of *we* in corpus B it is evident that this personal pronoun takes on a more aggregating function:

(15) *We need to go further*, through practical steps to reinforce social solidarity and stability, improving education and allowing human skills to flourish. (Europe for women. 2009)

(16) Each and every one of us will be given the opportunity to decide how *we want to* shape Europe. (Europe and you in 2008. A snapshot of EU achievements. 2008)

In the above examples *we* is used to express common necessity ('we need to') and aspiration ('we want to'), sentiments which unite people sharing the same needs and goals/expectations.

4.3. The EU as a rights-based union

According to Eriksen's (2009) third strategy of polity formation, i.e. the EU as a rights-based union, the institutions are seen as shaping their legitimacy through political means and the enforcement of a series of rights. As Eriksen (2009: 229) suggests, the EU is a political system that legislates and (re)allocates rights throughout Europe and whose actions impinge on identities and loyalties. A sense of community allegiance is thus derived from the transformation of citizens into right-holders who are cognizant of their precise entitlements. In the European Union, political responses to economic integration have produced an extension of rights. This extension, which might not have occurred independently among a collectivity of national states, creates a new basis for ensuring that politicians and procedures at all levels respond to citizens' demands. Expanded legal rights, just like electoral reforms or easier access to judicial institutions, are believed to open up more democratic opportunities for individuals and new routes of accountability (Chicowski 2006).

Scholars and experts give particular importance to rights when discussing accountability and aspects of democratic governance. Caporaso (2001: 10) believes that rights adhere to citizenship and occupy a prominent place in transnational democracy: "To be a citizen is to be a recognized member of a specific political organization and to exercise rights and responsibilities as a result of such membership". Rights also have an extremely strong solidarity-forming function. The EU brings civic rights into play as a way to spur participatory democracy, empower citizens and confer upon them a sense of identification with the institutions. When speaking about the activation of rights of citizenship, Habermas (1996: 506) interestingly defines them the 'juridification of new forms of inclusion', as rights imply an exercise on the part of the citizens rather than mere observance, and therefore an act of direct involvement.

From a discourse perspective, in its informative booklets not only does the EU foreground citizens' rights, but – most importantly – it highlights the fact that they are part of a harmonised and pan-European legislative framework set up to protect and safeguard citizens across the whole of the EU. Citizens' rights are indeed not geographically-bound but 'Europeanized': regardless of where citizens are, work or live within the Union, their rights are equally recognised and guaranteed.

In the whole corpus under examination, citizens are reassured they can claim the same rights across the EU by means of *wh-* determiners indicating indefiniteness, such as *whichever*, *whatever* and *wherever*:

(17) That is why the EU strives to ensure that *whatever* you decide to buy, and *wherever* you are in the EU, you are protected by key basic principles of consumer protection. (Your rights as a consumer. How the European Union protects your interests. 2007)

(18) EU citizens are free to live and work in *whichever* EU country they choose, so they should have equal access to civil justice everywhere in the European Union. (How the European Union works. Your guide to the EU institutions. 2007)

The formulations in the corpus concerning EU rights tend to be all-encompassing and cover all the possible options and possibilities. This is achieved through the use of: binomials (i.e. *live and work*, as in example 20); inclusive *or*; and conjunction *whether*; these elements are used to introduce all possible scenarios and contribute to making EU rights as all-inclusive as possible (example 19).

(19) Your rights are protected and you have access to redress *wherever you are* in the EU and *whether* you buy your goods in a shop, by mail-order *or* via the telephone and Internet. (Your rights as a consumer. How the European Union protects your interests. 2007)

(20) EU citizens are free to *live and work* in whichever EU country they choose, so they should have equal access to civil justice everywhere in the European Union. (How the European Union works. Your guide to the EU institutions. 2007)

As far as the longitudinal analysis is concerned, a comparison between the two subcorpora brings light to an interesting discursive change which takes place after June 2005. The adverb expressing spatial relation *anywhere* displays a significant increase in its number of occurrences in corpus B (Table 10), thus emphasising the lack of territorial restrictions of EU rights and stressing the concept of openness and freedom for citizens at transnational level.

	Corpus A	Corpus B
Anywhere	27	*53*
anywhere in the EU	6	*24*
anywhere in the Union	3	*4*
anywhere in Europe	3	*8*
anywhere in the European Union	2	*5*

Table 10. Occurrences and concordances of *anywhere* in both subcorpora.

In particular, Table 10 shows that the occurrence of the cluster *anywhere in the EU* increases by a factor of 4 after 2005. EU's accountability is therefore being increasingly grounded on the existence of a European-level set of rights which fosters a sense of allegiance among citizens and which is conducive to a greater sense of identification with the institutions (Fossum 2003: 336). It is in fact through space adverbials that the EU creates the image of a border transcending set of entitlements in an attempt to counteract the lack of a common identity background among EU citizens, supposedly at the root of the current democratic deficit (Caliendo/Napolitano 2010: 474).

5. Conclusion

In this chapter I looked at some suggested paths of institutional development through which the EU may construct its new identity role and which Eriksen (2009) identifies with that of: the problem-solver; the provider of benefits; and the provider of rights. The emergence of these new functional roles is related to processes of identity change on the part of the institutions and to subsequent transformations in the way citizens perceive the EU as a political entity. If in the eyes of the people the EU does not yet stand as a legitimate entity, it should try and portray itself as such and provide grounds for its accountability. After all, every time a new political community is created, as Bruter (2004: 186) suggests, the legitimacy of the contract that links it to the citizens and gives it institutional acceptability requires the creation of a new political identity.

As the analysis has shown, the EU is increasingly constructing its narrative and discursive self-representation by putting forward the image of an active agent and an effective problem solver. EU institutions are gradually moving away from the declamatory rhetoric of the past and increasingly appealing to tangible elements of identification in order to spur popular involvement and fill citizens' motivational vacuum. In such a perspective, the analysis has shown that the EU is mainly communicating with its citizens by promoting the benefits, opportunities and practical results that EU membership brings about (Calhoun 2003) as a way to enter into direct contact with them and become increasingly relevant to their daily lives (Risse 2004).

The evolution of EU discursive practices also works in the direction of constructing a 'we-group' which shares similar views, and a set of common values as well as a sound framework of rights. The rights attributed to citizens tend to acquire increasing relevance in the process of uniting members of the diverse and multicultural EU society within a common space. Foregrounding objective interests and establishing a common framework of entitlements, which are not geographically- or culturally-bound, are means of instilling a greater sense of solidarity within a community of individuals brought together by representation and participation. The process of creating a common EU civil law based on the notion of European citizenship, regardless of nationality, is crucial in this respect: "If being European citizens has to have a meaning, that meaning cannot be other than the opportunity to enjoy the same constitutional rights and fundamental freedoms, in whatever Member State citizens are." (Bisogni/Caroccia 2005: 207).[6]

References

Bisogni, Giacinto / Caroccia, Francesca 2005. La Cooperazione giudiziaria e il Diritto Civile. In Amato, Giuliano / Paciotti, Elena (eds) *Verso l'Europa dei diritti. Lo spazio europeo di libertà, sicurezza e giustizia.* Bologna: Il Mulino, 205-239.

Bruter, Michael 2004. Civic and Cultural Components of a European Identity: a Pilot Model of Measurement of Citizens' levels of European Identity. In Hermann, Richard K. / Risse, Thomas / Brewer, Marilynn B. (eds) *Transnational Identities. Becoming European in the EU.* Lanham: Rowman & Littlefield Publishers, 186-213.

Calhoun, Craig 2003. The Democratic Integration of Europe. Interests, Identity and the Public Sphere. In Berezin, Mabel M. / Schain, Martin A. (eds) *Europe Without Borders: Remapping Territory, Citizenship, and Identity in a Transnational Age.* Baltrimore: John Hopkins University Press, 243-274.

Caliendo, Giuditta / Napolitano, Antonella 2010. Consumers at the Heart of Europe: Language Practices and Identity Building Strategies. In Duszak, Anna / House, Juliane / Kumięga, Łukasz (eds) *Globalization, Discourse, Media: In a Critical Perspective / Globalisierung, Diskurse, Medien: eine kritische Perspektive.* Warsaw: Warsaw University Press, 443-486.

Caporaso, James A. 2001. Citizenship and Equality: A Long and Winding Road. Paper prepared for a theme panel at the Interna-

6 Original quote: "*Se essere cittadini europei deve avere un significato, tale significato non può essere che la possibilità di godere degli stessi diritti e delle stesse libertà fondamentali costituzionalmente garantite, in qualsiasi Stato membro ci si trovi*". [my translation]

tional Studies Association Annual Meeting, February 22-25, 2001, Chicago, Illinois.

Caporaso, James A. / Kim, Min-hyung 2008. The dual nature of European Identity: Subjective awareness and coherence. *Journal of European Public Policy* 16/1, 19-42.

Chicowski, Rachel A. 2006 Courts, Rights, and Democratic Participation. *Comparative Political Studies* 39/1, 50-75.

Citrin, Jack / Sides, John 2004. More than Nationals: How identity Choice Matters the New Europe. In Hermann, Richard K. / Risse, Thomas / Brewer, Marilynn B. (eds) *Transnational identities: becoming European in the EU.* Lanham: Rowman & Littlefield, 161-185.

Eriksen, Erik O. 2009. *The Unfinished Democratization of Europe.* Oxford: Oxford University Press.

Etzioni, Amitai 2007. The Community Deficit. *Journal of Common Market Studies* Vol. 45/1, 23-42.

Fairclough, Norman 1992. *Discourse and Social Change.* Cambridge: Polity Press.

Fossum, John Erik 2003. The European Union in search of an identity. *European Journal of Political Theory 2/3, 319-340.*

Goldsmith, M. 2003. Variable Geometry, Multilevel Governance: European Integration and Subnational Governance. In Featherstone, Kevin / Radaelli, Claudio M. (eds) *The Politics of Europeanization.* Oxford: Oxford University Press, 112-133.

Habermas, Jürgen 1996. *Between Facts and Norms. Contributions to a Discourse Theory of Law and Democracy* (translated by William Rehg). Cambride, Massachussets: the MIT Press.

Hermann, Richard K. / Brewer, Marilynn B. 2004. *Identities and Institutions: Becoming European in the EU.* In Hermann, Richard K. / Risse, Thomas / Brewer Marilynn B (eds) *Transnational Identities: Becoming European in the EU.* Lanham: Rowman & Littlefield, 1-24.

Keohane, Robert O. / Nye, Joseph S. Jr 2003. Redifining Accountability for Global Governance. Kahler, Miles / Lake, David A. (eds) *Governance in a Global Economy: Political Authority in Transition.* Princeton: Princeton University Press, 361-385.

Risse, Thomas 2004. European Institutions and Identity Change: What have We Learned? In Hermann, Richard K. / Risse, Thomas / Brewer, Marilynn B. (eds) *Transnational Identities. Becoming European in the EU.* Lanham: Rowman & Littlefield, 247-271.

Salvi, Rita (forthcoming). Exploring Political and Banking Language for Institutional Purposes. Gotti, Maurizio / Giannoni, Davide (eds) *Proceeding of the CERLIS Conference 2012*. Bergamo.

Schmitter, Philippe C. 1996. Examining the Present Euro-Polity with the Help of Past Theories. In Marks, Gary / Scharpf, Fritz W. / Schmitter, Philippe C. / Streeck Wolfgang (eds) *Governance in the European Union*. London: Sage, 1-14.

Scott, Mike 2008. *WordSmith Tools version 5*. Liverpool: Lexical Analysis Software.

Streeck, Wolfgang 1995. From Market Making to State Building? Reflections on the Political Economy of European Social Policy. In Leibfried, Stephan / Pierson, Paul (eds) *European Social Policy: Between Fragmentation and Integration.* Washington, D.C.: Brookings Institution, 389-431.

Taylor, Paul 2008. *The end of European Integration. Anti-Europeanism Examined.* London and New York: Routledge/UACES Contemporary European Studies.

Official documents (Last accessed: October 2012)

2005 Communication to the Commission. Community Action Plan to improve communicating Europe by the Commission. Available at: <http://ec.europa.eu/dgs/communication/pdf/communication_com_en.pdf>.

2005 Communication from the Commission to the Council, the European Parliament, the European Economic and Social committee and the Committee of the Regions - The Commission's contribution to the period of reflection and beyond - Plan-D for Democracy, Dialogue and Debate. Available at: <http://eur-lex.europa.eu/smartapi/cgi/sga_doc?smartapi!celexplus!prod!D

ocNer&lg=en&type_doc=COMfinal&an_doc=2005&nu_doc=494>.

2006 White Paper on a European Communication Policy. Available at: <http://europa.eu/documents/comm/white_papers/pdf/com2006_35_en.pdf>

2007 Communication to the Commission. Communicating about Europe via the Internet to engage the citizens. Available at: <http://ec.europa.eu/dgs/communication/pdf/internet-strategy_en.pdf>.

2007 Communication from the Commission to the European Parliament, the Council, the European Economic and Social Committee and the Committee of the Regions. Communicating Europe in Partnership. Available at: <http://eur-lex.europa.eu/LexUriServ/site/en/com/2007/com2007_0568en01.pdf>.

2008 Debate Europe. Available at: <http://ec.europa.eu/archives/debateeurope/index_en.htm>.

2008 Communication to the Commission. Communicating about Europe via the Internet. Engaging the citizens. Available at: <http://ec.europa.eu/dgs/communication/pdf/internet-strategy_en.pdf.>

2008 Communicating Europe through audiovisual media. Available at: <http://ec.europa.eu/dgs/communication/pdf/comm_audiovisual_media_en.pdf>.

2009 Evaluation of the Plan D / Debate Europe citizen consultation projects. Final Report. Available at: <http://ec.europa.eu/dgs/communication/pdf/evaluation_citizens_consultations_sep2009_en.pdf>.

Appendix

EU informative booklets edited and re-published after June 2005

It's a better life (2002)	Better off in Europe (Nov 2005)
Serving the people of Europe. How the European Commission works (2002)	Serving the people of Europe. What the Commission does for you (July 2005)
Choices for a greener future (2002)	A quality environment (Oct 2005)
Travelling in Europe (2002)	Travelling in Europe (2006, 2007, 2008, 2010, 2011-12)
Europe in 12 lessons (2003)	Europe in 12 lessons (2006, 2010)
Panorama of the European Union (2003)	Panorama of the European Union (2007, 2009)
How the European Union works. A citizens' guide to the EU institutions (2003)	How the European Union works. Your guide to the EU institutions (2007)
Many tongues, one family (2004)	Speaking for Europe. Languages in the European Union (2008)
Key facts and figures about Europe and the Europeans (May 2005)	Key facts and figures about Europe and the Europeans (2007)
Panorama of the European Union (Jan 2005)	Panorama of the European Union (2007, 2009)

Chiara Prosperi Porta

"Unity in Diversity": The Shaping of Identities in the Annual Reports of the ECB and EU Member States' Central Banks

1. Introduction

The accessibility of the web has contributed considerably to changes in the patterns of institutional communication within the EU, as well as to favour the circulation of 'replicated' online genres, that is to say the online version of a printed document (Petroni 2011) among Member States. The public dissemination of online official documents in the European communicative space can be considered a complex network, resulting from the overlapping of different national, regional and cultural backgrounds. This diversity variously shapes the concept of 'Europeanisation' (Papathanassopoulos/Negrine 2011) and does not render a homogeneous European identity, although over time this idea has been associated with the identification of an 'imagined community'(Sassatelli 2002). Consequently, the idea of a 'European discourse' springs from the multiplicity of the national politico-economic discourses (Schlesinger 1999), whose elite discourse community shares the same communicative space but can differently shape writing to disseminate knowledge. These dynamics are revealed in the genre of financial reporting within the EU.

Indeed, in recent years, much research has focused on how the notion of cultural orientation (Hofstede 2001; Katan 2004) can affect discourses, genres and frame identity (Hyland 1998; Simon 2004; Loseke 2007; Spencer-Oatey 2007; Samson 2010). These insights have particular relevance in the case of EU informative institutional documentation (Prosperi Porta 2009[1]; Salvi 2013) and the consolidated versions of related legal texts (Ardizzone/Pennisi 2010).

Some scholars (Bhatia 2004, 2008; Samson 2010) have also identified forms of discoursal hybridity during the textual release, as well as intertextual relationships (Frow 1991; Fairclough 1992, 2003; Bhatia 2010) within a shared communicative space. According to Frow (1991: 45)

> Texts are [...] traces and tracing of otherness. They are shaped by the repetition and the transformation of other textual structures [...]. The form of representation of intertextual structures ranges from the explicit to the implicit. In addition, these structures may be highly particular or highly general; they may be of the order of the message or of the order of the code. Texts are made out of cultural and ideological norms, out of the construction of genre; out of styles and idioms embedded in the language; out of connotations and collocative sets; out of clichés, formulae or proverbs; and out of other texts.

Given the dissemination of online financial documents in the EU and considering the impact of intertextuality on written discourse in a specific genre, it becomes possible to investigate the features of bank annual reports within the macro-genre of company reports in general because "exemplars of a genre exhibit various patterns of similarity in terms of structure, style, content and intended audience" (Swales 1990: 58).

More specifically, a few studies have concentrated on the role of business narrative and financial reporting (Garzone 2005; Bowker 2007; Wawra 2007; Malavasi 2010; Breeze 2012), explaining how discursive shaping can be affected by the distinctive feature of evaluation when using written English in the international financial world. As a matter of fact, in this sector, evaluative language is used when a positive image of financial performance is needed, thus conferring both an informative and a promotional nature to the text.

---

1 The construction of institutional discourse in these documents (White Papers and Standard Eurobarometer documents) is marked by typical linguistic features, traceable in *keywords* and *fixed collocations*, which contribute to developing identifying structures and patterns in the specific context of reference.

It has also been highlighted how narrative and legitimation techniques achieve an emotive and social force (Van Dijk 1998; Verschueren 2011) and are adopted by the authorship in order to engage in solidarity with the audience, and/or to restore confidence in the company/institution. In the light of the afore-mentioned considerations, the aim of this study is to analyse the Annual Reports of a selection of European Member States, comparing them with the same type of documents issued by the European Central Bank, starting on the verge of the financial shock in 2008 and investigating their evolution through to the recession in 2011. The paper therefore investigates convergent and/or divergent expressions of identity through the discursive representation of the common financial disaster affecting Europe over these four years, either at a national or supranational level.

In particular, the study analyzes the linguistic strategies used in this genre to shape identity and build public consensus, as well as the relevant keywords and phraseology. It also considers the features of narrative, evaluation, figurative language and possible intertextual elements involved in the realization of credible financial identities.

2. Corpus and methodology

The corpus is composed of documents issued by national central banks in the time span of the four years cited above, which have proved to be crucial for the economy in the Euro-zone. Four parallel sub-corpora were then obtained. The rationale behind the choice of these three specific member states (Greece, United Kingdom, Italy) compared to the ECB is to see how these countries have been affected differently by the crisis and to indicate their stance towards institutional measures and guidelines at a European level. These are traceable in their respective discursive identities and are marked in the language which they foreground and prioritize.

The sub-corpora were processed using the software ConcApp (version 5) as regards relevant word statistics, phraseology and collocations, both from a diachronic and synchronic perspective. The table below presents the characteristics of the four sub-corpora; each document is classified according to year, corpus size and country of reference.

country	2008 total words	2009 total words	2010 total words	2011 total words
ECB	139,253	127,619	127,352	93,769
Greece	104,418	26,729	22,111	19,587
Italy	118,783	114,565	109,322	109,318
UK	28,563	45,230	38,443	23,998

Table 1. Size of the sub-corpora.

The quantitative analysis will be combined with discourse analysis, focusing on the sections containing the Governor's foreword, in order to highlight the relevant discursive strategies involved in promoting the image of the institution and its policies.

3. Identity in the European communicative space: the shaping of discourse in the Central Banks' Annual Reports

The relationship between language and culture influences speakers' world-views and shapes identity (Scollon/Wong Scollon 2001). As argued by Heller (2011), language maintains or modifies the boundaries on which social organization and identity are based. If we consider the role of English as a *lingua franca* in the European financial context, it should also be questioned whether its communicative function is really detached "from the notion of a bond between a state, national identity or culture" (Narkunas 1999). In fact, the need to harmonise the cultural environment has represented a concrete objective in EU *institution-centred* communication, in view of promoting the sense of a common identity when orienting the Member States to share knowledge from different cultural perspectives. EU financial bodies are considered responsible for informing citizens about public policies, especially those which have an immediate impact on their lives. Therefore, identity needs to be built both into the single national and supra-national contexts.

A closer examination of each single report will investigate if a dishomogeneous perception of a European identity, a decline in the trust in Europeanisation and adverse economic conditions may affect institutional communicative patterns.

3.1. The ECB' s Annual Reports

The ECB's attitude towards the economy in Europe has been consolidated in the texts through cautious lexical choices, although evaluative and legitimizing language features are frequently deployed, in order to positively sustain the Bank's economic policies. Indeed, the underlying idea in all these reports is the sharing of common values and principles as a mandatory feature to achieve financial stability and institutional strength at a European level.

Some annual reports present a complex discursive shaping, sometimes showing a connection between the previous reports and the later texts, as an ongoing dialogue between the institution and the reader. In fact, over the four years investigated, the ECB's discourse has progressed from an initial attempt to promote institutional credibility and restore financial reputation, with the frequent use of words expressing commitment as well as public recognition, to a growing practice of lessening the downside recessional effects and forecasting a likely recovery. The following bar chart shows the seven most frequent words in the ECB's texts:

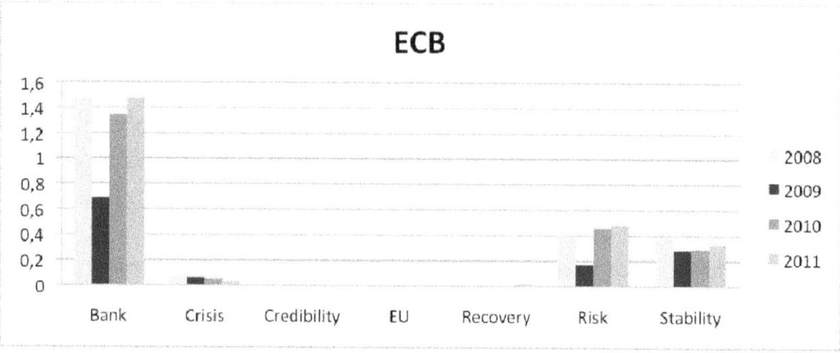

Figure 1. Keywords in the ECB's annual reports.

The active role of the *Bank* is noticeable, compared to low mention of *crisis*; the idea of *stability* as opposed to *risk* is juxtaposed almost equally, in order to appeal to the authority of the ECB as an institutional organization, and as a consequence, legitimizing its measures. On the contrary, other words have shown a different performance. In particular, the word *credibility* registers a very low occurrence in these reports, discouraging reflection on its own competence, as well as *EU* and *recovery*.

(1) The monetary analysis confirmed the assessment that *upside risks* to price *stability* prevailed. (ECB 2008: 10)

(2) […] a remarkable success in the light of adverse developments and a sign of the high *degree of credibility* of the ECB' s monetary policy. (ECB 2011: 8)

(3) The policy and regulatory responses to the financial crisis gathered momentum in 2009 with a number of concrete proposals for the enhancement of the *financial stability* framework at a European and global level. (ECB 2009: 11)

The most feasible reason for these specific lexical occurrences, as well as their different rating, can be two-fold: on the one side, there is the institutional discursive intent to set a balance between the aim of stability and the negative impact of fiscal measures; on the other, there is the intention to hide the financial imperfections that are affecting the Euro-zone, so as to positively build the ECB's identity.

There is also frequent use of evaluative adverbs in the texts, particularly those having a positive connotation, embodying the idea of accomplishment and deep involvement in duty, such as *actively, strongly, fully*. The adverb *actively* frequently occurs in two-word collocations, when discourse clearly aims at supporting the banking sector, thus emphasizing institutional role and commitment.

(4) The ECB *actively contributed* to the key policy and regulatory responses which were adopted in 2008 to address the financial crisis. (ECB 2008: 8)

The qualitative approach, targeted at unveiling the possible different realizations of identity, has revealed a wide use of narrative and legitimation strategies in the discourse. In the presentation of the crisis, most of the time, all the steps taken in the past by the ECB towards financial stability are described as significant achievements which soften the strong impact of the recession.

The Governor usually opens his presentation of the financial year, by introducing the element of narrative that helps his argumentation, as shown in the following examples.

(5) Following the severe intensification of financial market tensions in autumn 2008, the year 2009 started with a rapid and synchronised fall in economic activity worldwide. This was followed by a very gradual recovery in the course of the year, while inflation remained very low. Overall, euro area real GDP contracted by 4.0% in 2009 and average annual inflation stood at 0.3%. At the same time, medium to longer-term inflation expectations remained firmly anchored in line with the Governing Council's aim of keeping inflation rates below, but close to, 2% over the medium term. (ECB 2009: 10)

(6) In a coordinated move, the ECB and other major central banks announced reductions in their policy interest rates on 8 October 2008. Overall, between October 2008 and March 2009, the Governing Council lowered the key ECB interest rates by 275 basis points. The overnight money market rates fell even more steeply over this period, as the ECB significantly enlarged its liquidity provision to euro area banks. These unprecedented decisions were fully in line with the ECB's mandate to maintain price stability, in the light of declining inflationary pressures and risks. (ECB 2008: 6-7)

In order to rebuild or reinforce the Bank's credibility, often the texts show a voice of authority, performed by the use of evaluative modifiers that take the form of formulaic expressions (examples 7 to 10) so as to repeatedly mark this idea:

(7) Inflation expectations have remained [...] in line with price stability during this time, reflecting the *high degree of credibility* enjoyed by the ECB' s monetary policy. (ECB 2008: 10)

(8) This reflects the *high degree of credibility* of the ECB' s monetary policy. (ECB 2009: 10)

(9) This reflects favourably on the *high degree of credibility* of the ECB' s monetary policy. (ECB 2010: 10)

(10) [...] a sign of the *high degree of credibility* of the ECB' s monetary policy. (ECB 2011: 8)

Narrative and legitimation show their discursive importance, and are sometimes also highlighted in combination with metaphors (in the marine, liquid or organic domain). One popular metaphor is represented by the anchor of stability, which is used to express a solid identity in connection with the concepts of financial stability and public confidence; it manifests the ECB's reliability in the present-day difficulties of finance.

(11) In 2009 the European Central Bank *continued to operate as an anchor of stability* and *confidence* in the challenging environment created by the global financial crisis. (ECB 2009: 6)

In the organic metaphors group, the *illness/health* type is often present, especially in the 'contagion' form. This metaphor has been lexicalised in the language of economics and it represents a well-established model in financial literature (Prosperi Porta 2011: 328), achieving greater popularity in the last four years. Similarly, some intertextual features can be identified in the description of the Bank's attitude towards learning from the financial crisis, trying to justify the appropriateness of reforms for the sake of financial stability.

(12) The establishment of the European Systemic Risk Board (ESRB) on 16 December 2010 marked a major milestone in terms of Europe *learning lessons from the crisis.* (ECB 2010: 11)

As a matter of fact, the expression *learning lessons from the crisis,* not only refers to a lecture that was presented by the contemporary economist Stiglitz a year before this report, but it will also reveal its discursive relevance in other Member States' reports.

Metaphors can appear in collocations such as *economic recovery* and *sustainable growth* that juxtapose with *contagion effects* and *vulnerable public finances.* In the reports, the *illness* metaphor often appears in combination with liquid metaphors. In this way, the world finance is depicted as a dysfunctional set of organs in a human body and the excess of liquidity is described as drainage:

(13) Turning to structural policies, the ongoing *weakness* of economic activity and high levels of uncertainty about the economic outlook imply the need to *strengthen the resilience* of the euro area economy. (ECB 2008: 11)

(14) [...] the sustainability of public finances, especially in Greece, with *clear signs of contagion affecting* other euro area government bond markets. *Liquidity* in those segments of the euro secondary debt market, *dried up.* (ECB 2010: 10)

(15) *Sound* balance sheets, effective risk management and transparent, *robust* business models are key to *strengthening banks' resilience to shocks*, thereby laying the foundations for sustainable growth and financial *stability.* (ECB 2009: 11)

(16) This programme has enabled the Eurosystem to intervene [...] in the case of *dysfunctional market segments*. The *liquidity* effects *stemming from* these purchases have been fully neutralised by weekly *liquidity-absorbing* operations. (ECB 2010: 11)

These metaphors, as well as the use of personification, trace the diverse stages in the evolution from bad economic conditions to attendant institutional dedication to recovery, through the use of adjectives or nouns indicating health, strength and resistance.

The four texts show a convergent discursive shaping in the context of recasting reputation and in the stressing of the Bank's full commitment to the EU diverse socio-cultural environment, along with the importance of professional ethics in organisational issues. They can be considered in line with the use of positive terminology, figurative language and the mitigating discursive techniques exploited to dampen the negative effects of the crisis. These similarities can be traced in all the ECB's reports, although a new Governor has been appointed over the years.

3.2. The Bank of Greece's Annual Reports

Greece is the Member State that has been the most affected by the severe impact of the crisis and recession. The Annual Reports issued by the Bank in the four years under review present a different structure and length compared to the ECB documents. However, quantitative findings show that the same keywords also retrieved in the ECB texts, have a high frequency, as illustrated in the bar chart below:

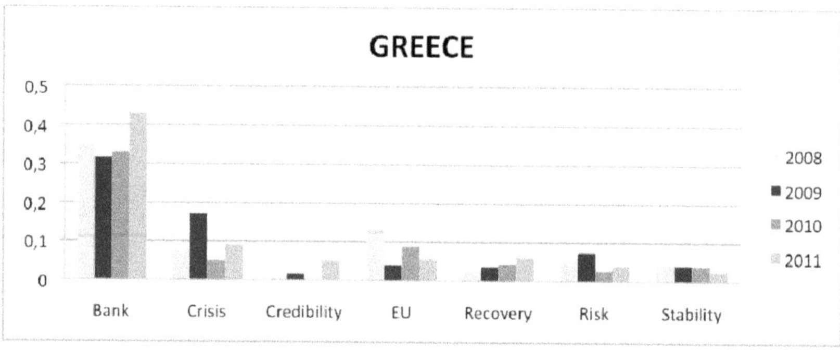

Figure 2. Keywords in the Bank of Greece's annual reports.

The mention of the *Bank* is high, as in the ECB data. However, unlike the ECB, reference to *crisis* is marked. The high occurrence of the word *crisis* (2008: 0.078%; 2009: 0.172%; 2010: 0.05%; 2011: 0.092%) is particularly relevant in the national perception of the current recession and its usage will equal only the Bank of Italy's reports. Reference to *credibility* (2008: 0.01%; 2009: 0.018%; 2010: 0%; 2011: 0.051%) is very evident at the height of the recession while overt mention of the *EU* fluctuates (2008: 0.132%; 2009: 0.041%; 2010: 0.089%; 2011: 0.056%). Instead *risk* and *stability* occupy relatively lower profile positions, *risk*, however, taking a more prominent place than *stability*.

From a lexico-pragmatic point of view, some words seem to have a high frequency when used in describing a careful institutional stance in very difficult times. In fact, lexis expressing caution, danger or a troubled situation occurs many times:

(17) In line with the *principle of prudence*, the Governing Council deemed it appropriate to establish a *buffer* of 5,736 million in total against *counterparty risks* in monetary policy operations. (Greece 2008: 13)

(18) Tapping into international savings, [...] is *difficult*, given the *negative climate* abroad concerning Greece and the prevailing *high uncertainties*. (Greece 2010: 10)

(19) Delays with fiscal adjustment and [...] *adverse conditions* still surrounding the provision of bank finance to the economy [...] are also contributing to *uncertainty* about the economic outlook. (Greece 2011: 10)

At times the authorship of the documents establishes intertextual connections with contemporary economic theory that is well known within the discourse community. Also in this country, as in the ECB's report, the text relies on the work of Stiglitz (2001), in this way contributing to the impression of an institution that is able to understand the public opinion's reserve:

(20) The causes of the crisis and effort required have not been adequately explained to the public. [...] A part of the public remains perplexed and reserved in the face of the unfolding changes, while there are also reactions that aim to revert the practices of the past. Owing to a significant information deficit, the public, though aware of the critical turning point we have reached, has yet to fully grasp the causes of the crisis, or the scope, duration and absolute necessity of the efforts required. [...] The new framework will require all Member States to make long-term commitment, especially in terms of fiscal discipline. (Greece 2010: 9-10)

The forewords published during the first critical years (2008-2009) already show a move towards a new type of structure that will be exploited in the latest documents (2010-2011). They present an introductory note explaining all the policies implemented during the last financial year and provide a narrative part that is embedded in the other sections, having the function of legitimizing the Bank's identity and acknowledging its institutional recognition:

(21) Throughout 2008, the Bank continued to strengthen its relations with central banks and banking supervisors of South-Eastern Europe, especially in the countries where Greek banks are present. (Greece 2008: 15)

Sometimes, the Governor uses some discursive strategies to mark national identity in terms of persuasion and space. In fact, the national bank's positioning in the European financial space marks its adherence to the ECB's guidelines, civic 'inclusiveness' and sense of belonging to the country's identity, as opposed to impersonal distance. This is revealed in the use of personal deixis, represented by pronominal usage, that describes the author's proximity to the nation but also to the banking institution:

(22) If [...] *one* adds the important cyclical problems caused by the recession, then it becomes clear that the difficult task before *us* calls for a persistent and coordinated effort over several years.[...] Failure to tackle the recession could compromise *our* ability to meet the targets of fiscal consolidation. [...] In the current context, growth requires the mobilisation of the private business sector; this cannot be achieved as long as the state continues to dominate the economy. (Greece 2011: 11)

Consequently, special emphasis is given to the importance of a credible institutional reputation. In order to successfully do this, the focus is on markers of value and persuasive lexis that index authorial solidarity, so as to accommodate the national identity in a space that is European, but detached from the State:

(23) In 2009 the Bank of Greece *successfully pursued its mandate* in all of its areas of responsibility. As in the past, this would not have been possible without the *commitment, dedication and competence* of the Bank's staff which [...] *worthily* represents the Bank on Eurosystem committees [...] and at other European and international fora. With its *consistent performance* in fulfilling its statutory obligations, including Eurosystem-related tasks, the Bank has won

itself *widespread recognition* and *a special place* in the country's institutional landscape [...]. (Greece 2009: 14-15)

(24) In the current context, growth requires the mobilisation of the private business sector; this cannot be achieved *as long as the state continue to dominate the economy.* (Greece 2011: 11)

Interestingly, the shared values of democracy and social responsibility come to the fore, particularly in the most recent forewords (2010-11) where the sacrifices made by the Greek people are publicly acknowledged and are used, along with figurative language, to help argumentation.

(25) [...] for the *sacrifices made by the Greek* people not to have been in vain, society must at this critical juncture take a *responsible stance* [...] so as *to win the war on all fronts*, starting with the construction of an efficient and flexible state that will serve both the competitive functioning of markets and social cohesion [...] This will avert conditions that would lead not only to *our sacrifices* going to be wasted, but also to a drastic deterioration in the standard of living. (Greece 2011: 15)

In fact, unlike the 2008 and 2009 forewords, the latest texts (2010-2011) show a more frequent display of figures of speech, particularly illness metaphors, as the crisis evolves and damages the national identity; the Greek economy is frequently described as being in a state of distress, aiming at full recovery and fighting against relapse.

(26) Work continued towards a *reinforced* framework of economic governance and policies to *prevent a recurrence of the crisis.* (Greece 2010: 9)

(27) Economic *recovery* therefore hinges upon the setting in motion of a *virtuous circle*, whereby an economic outlook fosters the *strengthening* of the banking system. (Greece 2011: 13)

Engineering and transport metaphors are displayed, also, in order to help the authorship to manifest optimism about the economic future, thus feeding positive expectations in the readership:

(28) Instead, the economy will be able to *get back on track* to recovery by the end of 2013 and gradually into a *virtuous circle* of confidence-fiscal balance-growth. (Greece 2011: 15)

3.3. The Bank of England's Annual Reports

The shaping of identity in the documents issued by the Bank of England results in a clear intent to show a strong institutional but independent image in three layers of discourse messaging: the Bank's identity is built up through a claim to shared European values, through its intent to communicate financial performance publicly at a national level, and through the effort to cooperate within a common European economic framework.

Each report contains two different, short forewords: a technical one by the Chairman of the Committee of non-Executive Directors, the other by the Governor. This is an evident mark of the double nature that connotes the Bank, both public and private, and of its independence from the State. In fact, the two forewords are the expression of two distinct identities conflated in the institutional one.

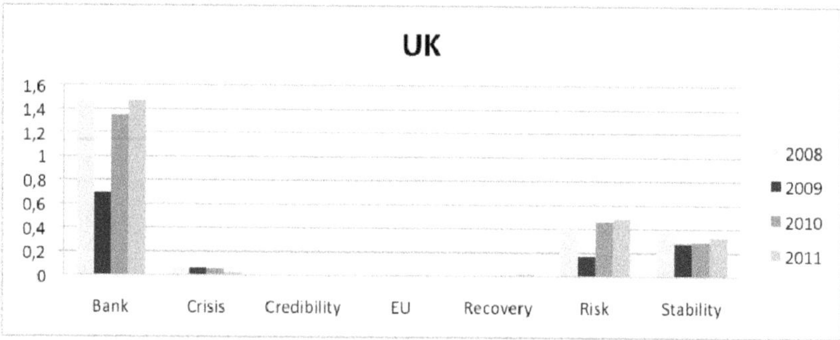

Figure 3. Keywords in the Bank of England's annual reports.

The textual analysis of frequencies has confirmed some analogies but also shown differences with the other countries. Reference to the *Bank* is very high, on *a par* with the ECB data, and significantly higher than that of the Bank of Greece; on the contrary, *crisis* is noticeably understated. Words like *credibility*, *EU*, and *recovery* are little mentioned, possibly reflecting the desire to communicate an image of a bank which is a well-established and healthy institution in no need of supranational financial endorsement. Also, similar to the ECB reports, *risk* and *stability* are fairly evenly juxtaposed: *risk* (2008: 0.4%; 2009: 0.172%; 2010: 0.459%; 2011: 0.487%), as one of the most representative in the texts, is a marked expression of the perceived danger and turmoil in the national financial space, but it is associated with nouns such as *avoidance, governance, management* and *tolerance,* all suggesting fundamental and efficacious control by the Central Bank authorities.

In all the four documents, narrative is used as an introduction, either appearing in the Chairman's or the Governor's foreword, as shown in the following examples:

(29) The year 2008/09 presented an unprecedented set of challenges for the Bank as it responded to the ongoing financial crisis and the significant economic downturn now affecting economies globally. (England 2009: 2)

(30) The past year has been the most challenging for the Bank of England since it gained independence in 1997. (England 2009: 4)

The Governor's forewords offer interesting examples of authorial positioning, by emphasizing dedication to resolving the difficult economic situation, as well as the institutional learning attitude towards the crisis, as already shown in the reports of the ECB.

(31) All parties *have learnt lessons from the period of financial turmoil.* (England 2008: 3)

(32) Like others, I am convinced that one of the *key lessons from the current financial crisis* is the need for some form of macroprudential and systemic regulatory regime. (England 2009: 3)

This institutional supportive intent is highlighted by evaluative patterns, in order to confer authority to the initiatives of the Bank and its executives. Evaluation is marked not only by the display of opinion verbs, but also by nouns and adjectives referring to the excellent banking performance and sense of duty:

(33) *I have witnessed* the *extraordinary* level of personal and professional *commitment* that has been maintained throughout this time. I *think* it is *fair* to say that as time has progressed the Bank's position has been increasingly understood as the true nature of the banking crisis has unfolded. (England 2008: 3)

Therefore, most forewords show authorial endorsement and construe a shared identity addressing discourse both to the community of practice and the general public:

(34) *I have made no secret* over the past couple of years of *my belief* that the path of the economy would soon become much bumpier than during the first ten years of the Bank of England's independence. [...] *we are now faced* for the second time in less than two years with the prospect of a sharp, but temporary, rise in inflation. (England 2008: 4)

Concepts like leadership, dignity and ethics are also highlighted, in order to legitimize authority.

(35) [...] The *commitment and dedication* of Bank staff has been *exemplary* throughout [...] *I should like to place on record my admiration* for the Governor who, under intense pressure, has provided *dignified and dedicated leadership* throughout. (England 2009: 3)

Sometimes the construction of identity can also be supported by metaphors, particularly the illness/health/war types, when it is necessary to make a clear distinction between national, institutional and European identity.

(36) The bank's problems were an *early symptom of a much more generalised malaise* in the dollar, sterling, euro and other financial markets that has proved deep and long lasting. [...] we faced a *deteriorated* situation in which banks' fears about each other were compounded by the impact of wild market rumours about perfectly *healthy institutions*. [...] The Bank of England is well known by its peers to be among *the leanest* of central banks. [...] That applies to the *front line operations* in markets, financial stability and banking and also to the rest of the Bank, where almost every activity has been drawn into the credit crisis in one way or another. (England 2008: 4-5)

National, institutional and European identities shape the image of the Bank. The first aspect that is exemplified in discourse is the Bank's independence from the State but also its dedicated contribution to the economy; secondly, the idea of a solid institutional reputation; thirdly, the concept of national identity, considered as one cohesive political group. These three different facets contribute to create institutional reliability.

The independent identity and distance of the Bank from current politics is often pointed out in the authorial inclusive positioning in the financial space, through the occurrence of personal deixis in debating fiscal policies, the use of repetition combined with pronominal usage in reiterating cohesive institutional action, and narrative, all going towards conveyance of a well-established national image:

(37) The Bank's *new statutory responsibility* for financial stability is welcome. But, as I have made clear, I regret that the *new responsibility* has not been accompanied by any new powers to deal with banks before they fail. [...]The past year has been one of great change at the top of the Bank. In February, Sir John Gieve stood down as Deputy Governor for Financial Stability. Nigel Jenkinson, after almost 32 years in the Bank, the last six of which as Executive Director for Financial Stability, is retiring. [...] I am grateful for their *independent views* and their *contribution* to the policy debate in what has been a challenging period. (England 2009: 5)

(38) I wrote an open letter to the Chancellor, the seventh since the Bank gained its independence thirteen years ago. [...] At the heart of this debate is the paradox that in our market economy we have banks that are 'too important to fail'. We cannot allow that anomaly to continue. (England 2010: 3)

(39) *We have* established asset purchases [...] *We have* continued to press the case for fundamental reform [...] And *we have* a new Chairman of the Court and a new governance structure. (England 2010: 3)

In the most recent forewords, the identity of the Bank which emerges is that of institutional commitment, enhanced by legitimation strategies. Institutional identity is associated with civic identity, sense of responsibility and continuous commitment on the part of all its members. Therefore, lexical choices collect around those keywords in the semantic category of accountability and stability:

(40) The PRA [Prudential Regulation Authority] will contribute to the *stability* of the UK financial system by assessing the *safety and soundness* of individual

> financial firms. It will build on the *improvements* to regulation [...] based more on *judgement* and less on *compliance*. Appropriate *accountability* for our performance in delivering financial *stability* will be a keystone of the new regulatory framework. *The Bank must be accountable to Parliament and the public.* (England 2011: 3)

The Governor shows a positive stance in his conclusions, his argumentation describing an optimistic view of the future challenges the Bank will have to face, and its commitment to steering the road to recovery:

(41) Finally, I am honoured to have been reappointed as Governor [...] and will try to the best of my abilities to rise to the varied challenges that will undoubtedly confront the Bank over the next five years. (England 2008: 5)

(42) I have no doubt these years will see their own challenges. [...] We are a richer, prouder institution. (England 2009: 5)

3.4. The Bank of Italy's Annual Reports

Unlike the structure of the previous reports, the Governor's statement comes at the end of the Bank of Italy Annual Report in the form of *Concluding Remarks*. In this part, the Governor's aim is to sum up all the issues examined in the entire report concerning the financial year and make comments on the Bank and State actions. The remarks are addressed to the Bank's shareholders and are publicly delivered during the ordinary meeting. In this respect, therefore, these texts are analogous with corporate shareholders' letters.

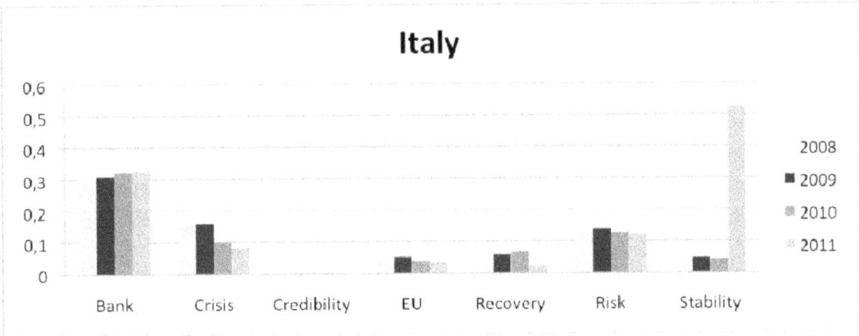

Figure 4. Keywords in the Bank of Italy's annual reports.

Reference to the national *bank* is considerably lower than in the UK data, while at the same time the *crisis* is fully acknowledged. A striking feature is the rocketing of the term *stability* in 2011 (0.527%), after a low occurrence (2008: 0.046%; 2009: 0.049%; 2010: 0.043%), as opposed to *crisis*, that performed higher in 2008 and 2009 than the latter two years. This is probably linked to the dramatic financial insecurity then affecting the country and to the authorial aim of mitigating this phenomenon in discourse, possibly representing a strategic claim for authority and institutional responsibility in gross financial dysfunctions.

A different trend is shown in the low values for *credibility* and the absence of statistically significant occurrences for *recovery*, indicative, possibly, of a lack of willingness to make forecasts or to attract overall judgements about the Bank's performance.

The texts are composed of an introduction that briefly presents both the plans for the future and a narrative review of the past financial year. Narrative is used for different reasons. First, it persuades the readership about the importance of financial reforms; secondly, it promotes the Bank's reliable reputation and positively supports the synergy between the State and finance; thirdly, it creates an inclusive identity, ending on an optimistic note. In order to shape these different aspects of identity, frequent emphasis is put on institutional commitment and civic solidarity, as shown in collocations such as *extraordinary commitment, social protection, professional and human qualities* of the Bank's staff. As a result, the Governor shapes discourse in order to offer a collective identity, often marked by the personal pronouns *we* and *I,* thus indexing the Bank's employees as a cohesive group of people working for the entire community.

(43) The response to the debt crisis is first and foremost in national policies, in the complete implementation of the adjustment plans that have been agreed. *Solidarity* among the member countries of the Union must be matched by a *sense of responsibility and compliance with the rules.* (Italy 2010: 8)

(44) Following the absorption of the Italian Foreign Exchange Office, for more than a year the Bank of Italy has had the function of combating money laundering through a Financial Intelligence Unit [...]. In carrying out its tasks, the FIU has developed fruitful synergies with the banking and financial supervision area. This benefits the stability and the reputation of the banking system. The Unit cooperates closely with the judiciary [...]. (Italy 2008: 1)

In this way, authority and a prestigious reputation are endorsed by the frequent mention of actions that are in accordance with the government's guidelines. Also, over the last few years, national and European identity have converged in order to emphasize institutional accountability. However, the will to achieve independence from political power is stressed in the *Remarks*:

(45) The Bank of Italy has been a precious source of dedicated public servants for Italy and Europe. Merit and independence: these are the essential conditions for the credibility of its analyses. (Italy 2010: 4)

Therefore, the Bank describes its mission in the rehabilitation of a tarnished institutional reputation, in order to win back civic credibility.

(46) When I first took the floor before this annual meeting [...] I said that I felt that one of my duties was to lead the Bank of Italy towards change [...], making the Bank once again the *trusted,* independent advisor to Parliament, the Government and the general public. (Italy 2010: 3)

(47) On 1 November last year Mario Draghi took up the post of President of the ECB. [...] During these years, his efforts, including in the demanding role of Chairman of the Financial Stability Board, brought *prestige to our institution and strengthened its reputation* both in Italy and abroad. [...] We will devote all our abilities, our utmost efforts, to overcoming them in the interest of the

Bank and of the country. [...] The women and men who work in the Bank at every level and in every field are its lifeblood and its strength. (Italy 2011: 3- 4)

Authorship appeals to uprightness and solid tradition, as intrinsic values belonging to the Italian banking system.

(48) In our country there was no banking crisis. [...] In line with international guidelines, we now have to revise the framework of rules [...]: giving the Bank of Italy, as supervisory authority, the power to remove corporate officers responsible for conduct harmful to the sound and prudent management of a bank. (Italy 2010: 17)

(49) The Bank of Italy is attentive to the need to ensure basic correctness in bank-customer relations. To this end, it has stepped up cooperation with the other authorities in the financial sector; (Italy 2011: 13)

At times, authorial positioning in the financial space also entails a criticism of inefficient civil justice and ineffective economic response which prevents growth, and includes an appeal to a united identity in a joint national and European political culture.

(50) Our analyses point the finger at Italy's productive structure, which is more fragmented and static than in other economies; and at government policies that fail to encourage, and often hamper, its development. (Italy 2010: 12)

(51) If governments, the EU authorities, the ECB itself, judge the progress of the troubled countries in financial restructuring and structural reform positively, this must be followed by a practical commitment on their part to orient the markets' assessments in the same direction. (Italy 2011: 15)

(52) The crisis has reminded us brutally of the importance of joint action, common objectives and policies, shared sacrifices. This lesson goes for the world, for Europe and for Italy. (Italy 2009: 14)

(53) Thanks to hard work, ingenuity and sacrifice, in the 150 years of its history as a nation Italy has made enormous progress in the material conditions of life. We have experienced periods of booming growth. During the first fifteen years of the twentieth century per capita gross domestic product increased by 30 per cent. [...] What unites us is more powerful than what divides us. Today, first of all the public budget must be restored to its proper role as a factor of stability and an engine of economic growth. (Italy 2010: 18-19)

As a result, weak economic performance and critical financial conditions are frequently displayed in medical metaphors. Market failure and stasis often take the form of contagion or of an invalidating disease:

(54) The market rejected economic policy interventions despite their timidity; *blinded*, it lost its diagnostic capacity, its self-correction mechanisms were *paralysed*. (Italy 2008: 5)

(55) The coming months will be decisive; excessive *mortality* due to financial *strangulation* of too many companies that have the potential to prosper after the crisis has passed is a second, serious risk for the Italian economy. (Italy 2008: 8)

(56) [...] the unavailability of a process permitting orderly management of the debt crises of sovereign states, amplified the *damage* and *contagion* and at the same time heightened moral hazard. (Italy 2009: 5)

Metaphors of health or recovery are helpful also when manifesting the urge to restore public confidence in finance.

(57) The *wound* that has been inflicted upon the public confidence [...] in the social compact that *binds us together – has to be healed*. (Italy 2008: 16)

(58) Italy has important tasks to perform. It has already begun to work on [...] a public sector that keeps its accounts in order, is not wasteful, and facilitates the economy; a *sound and efficient* banking system and a productive system that is capable of innovating, competing and growing. (Italy 2011: 16)

Military metaphors instead can be created to re-enforce the Bank's traditional role in pursuance of the common social benefit:

(59) In 1992 [...] The *struggle* was long drawn out: under a flexible exchange rate regime, three years on, spread still exceeded 650 basis points. Yet the *battle was won*, because [...] stability had been incorporated into the country's political culture. (Italy 2009: 15)

Sometimes both narrative and metaphors are employed in the last part of the *Remarks* when there is the need to drive argumentation towards an optimistic ending:

(60) Italian banks' balance sheets are not weighed down by a legacy of impaired assets. Let them [...] follow the example of bankers who financed the reconstruction and economic boom of the 1950s and 1960s. [...] Confidence cannot be rebuilt with false hopes. [...] Surely we can emerge from this crisis stronger than before. (Italy 2008: 17)

(61) Today's *battle* too – to combine fiscal discipline with the return to growth must be *fought* by calling on those same values that have enabled us, together, to *vanquish* the difficulties of the past: the capacity to act, equity; the desire to know, solidarity. (Italy 2009: 15)

At the same time, the concluding section expresses the need for cohesion between politics and economics, forecasting a better world which is tailored to the expectations of Italy's future generations:

(62) Italian society cannot avoid confronting a changed world in which no position rents are allowed. At the same time, politics must ensure the prospect of a profound renewal that nourishes hope and responds to the aspirations of the younger generations. (Italy 2011: 17)

4. Concluding Remarks

The analysis has revealed some common ground as well as important differences in terms of identity and space in the texts analysed. The discursive importance of the national cultural terrain in the European financial space is signalled by the use of legitimation and narrative strategies, along with figurative language.

Some specific keywords have shown a high occurrence in all the reports and over the four years. This has revealed a fairly homogenous perception of the crisis among the Member States, although the significant differences in the lexical rates according to the country of reference are a distinctive mark either of national or institutional identity. In fact, those countries (Greece and Italy) whose reputation has been seriously tarnished have shown a higher occurrence of several keywords, thus showing a strict relationship between the need to rehabilitate a deteriorated reputation and specific lexical choices.

This is reflected in the qualitative findings in terms of a more or less marked distant institutional positioning in the sub-corpora. In fact, on the one hand, the countries that have been badly affected by the crisis present a more visible authorial endorsement and proximity to the government's and the ECB's policies; on the other, the U.K. and the ECB, which have shown a stable financial performance in spite of the recessional impact, have shaped their discourse without the need to emphasize the idea of a credible trans-regional institutional reputation.

Not surprisingly, the ECB reports are the only ones that really aim at embodying a European and *Europeanised* identity, expressing institutional dedication at a supranational level and a harmonised approach to financial problems. On the contrary, the texts issued by the three member states reveal a lack of cohesive European identity, stressing their single national identities and space.

A comparison of the data also reveals their heterogeneous textual structure. Most reports contain a narrative introduction with the Governor's comments on the past and present financial crisis that shift from an optimistic general forecast of the incoming financial year to the legitimisation of the decisive action of the bank at that time, marking values such as diligence and willingness to repair recession and restore growth and public confidence. This is analogous with corporate CEO's letters to shareholders, where an optimistic ending usually concludes argumentation.

In some reports, authorship has built up intertextual connections among different texts over time, thus emphasizing their ongoing dialogic nature, engaging both the institution and its readership in the civic, financial space. Indeed, one avenue for future research could well lie in how institutional perceptions and their own textualizations are affected by expressions of public opinion in the different Member States in response to European and national economic policies.

References

Ardizzone, Patrizia / Pennisi, Giulia A. 2010. The Discursive Construction of European Identity: Stylistic Analysis of "Text in Context". *Textus* XXIII/1, 140-60.

Bhatia, Vijay K. 2004. *Worlds of Written Discourse*. London and New York: Continuum.

Bhatia, Vijay K. 2008. Genre Analysis, ESP and professional practice. *English for Specific Purposes* 27/2, 161-74.

Bhatia, Vijay K. 2010. Interdiscursivity in Professional Communication. *Discourse and Communication* 21/1, 32-50.

Bowker, Janet 2007. Uncovering intentionality and the process of persuasion in oral business narratives. In Bamford, Julia / Salvi, Rita (eds) *Business English: Language at Work*. Roma: Aracne, 47-89.

Breeze, Ruth 2012. Legitimation in Corporate Discourse: Oil Corporations after Deepwater Horizon. *Discourse & Society* 23/3, 1-17.

Fairclough, Norman 1992. *Discourse and Social Change*. Cambridge: Polity Press.

Fairclough, Norman 2003. *Analysing Discourse: Textual Analysis for Social Research*. London: Routledge.

Frow, John 1991. Intertextuality and Ontology. In Worton, Michael / Still, Judith (eds) *Intertextuality: Theories and Practice*. Manchester: Manchester University Press, 45-55.

Garzone, Giuliana 2005. Discorso e dinamiche linguistiche nella comunicazione professionale: il caso delle Lettere agli Azionisti. In Bellini, Giuseppe / Merlini, Lavinia / Vecchiato, Sara (eds) *Pablo Neruda in Bocconi: Uno sguardo alle lingue professionali*. Milano: Egea, 121-36.

Grieves, Chris 2005. ConcApp 5 software tool. <http://www.edict.com.uk/PUB/concapp/>.

Heller, Monica 2011. *Paths to Post-nationalism: A Critical* Ethnography *of Language and Identity*. Oxford: Oxford University Press.

Hofstede, Geerte 2001. *Cultures' Consequences: Comparing Values, Behaviours, Institutions and Organisations across Nations* (2nd ed.). Thousands Oaks (CA): Sage Publications.

Hyland, Ken 1998. Exploring corporate rhetoric: Metadiscourse in the CEO's letter. *Journal of Business Communication* 35/2, 224-45.

Katan, David 2004. *Translating Cultures*. Manchester: St. Jerome.

Loseke, Doneleen 2007. The study of identity as cultural, institutional, organizational and personal narratives: Theoretical and empirical integrations. *The Sociological Quarterly* 48/4, 661-88.

Malavasi, Donatella 2010. The multifaceted nature of banks' annual reports as informative, promotional and corporate communication practices. In Evangelisti Allori, Paola / Garzone, Giuliana (eds) *Discourse, Identities and Genres in Corporate Communication: Sponsorship, Advertising and Organizational Communication.* Bern: Peter Lang, 211-34.

Narkunas, Jean Paul 1999. *The Cash-Nexus of Language: the World Bank and the Study of English,* available on line: <www.gradnet.de/pomo2//pomo2.papers/narkunas00.htm>.

Papathanassopoulos, Stylianos / Negrine, Ralph 2011. *European Media*. Cambridge: Polity Press.

Petroni, Sandra 2011. *Language in the Multimodal Web Domain.* Roma: Aracne.

Prosperi Porta, Chiara 2009. Towards Sustainable European Integration. In Salvi, Rita (ed.) *L'insegnamento delle lingue in Italia in relazione alla politica linguistica dell'Unione Europea,* working Paper 59/2009 del Dipartimento di Studi Geoeconomici, Linguistici, Statistici e Storici per l'analisi regionale. Roma: CNR digital library, <http://150.176.3.132/990>, 48-56.

Prosperi Porta, Chiara 2011. The metaphor of illness and health in economics. In Di Martino, Gabriella / Lombardo, Linda / Nuccorini, Stefania (eds) *Papers from the 24th AIA Conference, Challenges for the 21st Century: Dilemmas, Ambiguities, Directions.* vol. 2. Roma: Edizioni Q, 321-31.

Salvi, Rita 2013 (forthcoming). Exploring Political and Banking Language for Institutional Purposes. *Proceedings of the CERLIS Conference 2012.* Bergamo.

Samson, Christina 2010. Boosting the Company's Identity. Point of View and Promotional Discourse Style in a Corpus of Business to Consumer (B2C) Webpages. *Textus* XXIII/1, 139-60.

Sassatelli, Monica 2002. Imagined Europe: the shaping of a European cultural identity through EU cultural policy. *European Journal of Social Theory* 5/4, 435-51.

Scollon, Ron / Wong Scollon, Suzanne 2001. *Intercultural Communication.* Malden Mass.: Blackwell.

Schlesinger, Philip 1999. Changing spaces of political communication: the case of the European Union. *Political Communication* 16, 263-79.

Simon, Bernd 2004. *Identity in Modern Society. A Social Psychological Perspective.* Oxford: Blackwell.

Spencer-Oatey, Helen 2007. Theories of identity and the analysis of face. *Journal of Pragmatics* 39, 639-56.

Stiglitz, Joseph 2001. *Information and the Change in Economics.* Nobel Prize Lecture: Stockholm, available on line: <www.nobelprize.org>.

Stiglitz, Joseph 2010. Lessons from the Global Financial Crisis of 2008. Revised version of a lecture presented at Seoul National University on October 27, 2009. *Seoul Journal of Economics* 23/3. Seoul: Institute of Economic Research, 321-39.

Swales, John 1990. *Genre analysis. English in Academic and Research Settings.* Cambridge: Cambridge University Press.

Van Dijk, Teun 1998. *Ideology: a Multidisciplinary Approach.* London: Sage.

Verschueren, Jan 2011. *Ideology in Language Use: Pragmatic Guidelines for Empirical Research.* Cambridge: Cambridge University Press.

Wawra, Daniela 2007. On Course for the Next Stage of Success: The Annual Reports of U.S. and Japanese Companies. In Garzone, Giuliana / Ilie, Cornelia (eds) *The Use of English in Institutional and Business Settings.* Bern: Peter Lang, 127-46.

Web references

European Central Bank:
<http://www.ecb.int/pub/annual/html/index.en.html>
Bank of England:
<http://www.bankofengland.co.uk/publications/Pages/annualreport/default.aspx>
Bank of Greece:
<http://www.bankofgreece.gr/Pages/en/Publications/GovReport.aspx>
Bank of Italy:
<http://www.bancaditalia.it/pubblicazioni/relann>

LAURA FERRAROTTI

Promoting Identity in Italian University Websites

1. Introduction

This chapter discusses the results of a linguistic analysis of the English version of sixty Italian university websites focusing specifically on how the personal pronouns *we* and *you*, and the possessive adjectives *our* and *your* are used in the introductory sections of these websites. It examines, in particular, the various ways such pronouns are used by these universities to present themselves, and how they have been deployed to appeal to prospective international students through the construction of a strong and credible institutional identity. This study considers how these universities' self-promotion strategies are closely aligned to the way corporations promote their corporate strategies and decisions.

The mainly promotional stance adopted by both private and public institutions has been analyzed in numerous linguistic studies (Cheney 1983; Swales/Rogers 1995; Hyland 1998; Medina 2005; Fairclough 2006; Dewey 2007; Salvi/Bamford 2007; Caiazzo 2009, 2010, 2011). The main assumptions underpinning most of these studies consider how the impetus given to the marketization of public institutions, along with the privatization of higher education (Kuong 2000; Altbach 2004), has greatly contributed to a profusion of promotional texts that, in the case of the university, go well beyond the mere informative descriptions previously found in university brochures. Genres ranging from printed fliers, pamphlets, prospectuses and posters, on the one hand, to virtual websites, on the other, are just a few examples of texts which have recently taken on a conspicuously promotional flavor, in many ways analogous to the way businesses advertise their products (Askehave 2007; Osman 2008). It follows, therefore, that the so-called marketization of academic discourse (Fairclough 1993) has been influenced by the ongoing globalization of economic markets, and that this has led to increased competition among universities.

Linguistic choices have been inevitably influenced by how universities use language to rebrand and promote themselves (Osman 2008). Such strategies have been formulated to respond to the changing relationship between universities and students. Fairclough (1993) pointed out that stable and authoritative universities were previously in a position to select their students, whereas now students actually choose universities as if they were commodities. Thus, students are viewed as consumers who choose the university that best suits their study, social and career needs (Wernick 1991; Fairclough 1993; Benwell/Stokoe 2006). The approach adopted by universities to introduce and promote themselves by using strategic personal pronouns in their brochures and websites contribute to achieving these objectives. These features have actually become determining factors in re-branding, projecting, reinforcing and reaffirming a strong corporate identity for Italian universities.

Universities deliberately use personal pronouns and possessive adjectives, *we*, *you*, *our* and *your*, for self-promotional introductory texts. The universities strategically either include or exclude their audience, albeit wholly or partly, in order to project an authoritative image. Moreover, apart from constructing and maintaining a credible institutional identity, the use of pronouns, especially of *we* inclusive and *you*, also serve to enhance the proximity and involvement of the audience.

Studies abound about the use of pronouns and how they have been employed in various texts such as university lectures, academic texts, political discourse, TV news and reports, university brochures and for the variety of target audiences they address - academics, professionals, potential students, experts (Rounds 1987; Hyland, 1998, 2002; Fortanet, 2004; Walsh, 2004; Crawford Camiciottoli, 2004, 2005; Ferrarotti, 2009; Caiazzo, 2010, 2011). The majority of these studies points out that what exactly a pronoun actually refers to can sometimes be problematic and difficult to discern. As Biber *et al* say: "Most typically, personal pronouns are used to refer to definite specific individuals [...]. However, the specific reference is often far from straightforward" (Biber *et al* 1999: 328). In fact, pronouns may vary in function across a particular text, thereby generating ambiguity. For example, the pronoun *we* "usually refers to the speaker/writer and the addressee (the inclusive *we*), or to the speaker/writer and some other person or persons associated with him/her (exclusive *we*) and not the addressee" (Biber *et al* 1999: 328).

Finally, the university websites analyzed in this study show that the pronouns *we* and *you* and the possessive adjectives *our* and *your*, have the underlying aim of contributing to the construction of the university's identity with prospective students. How this is achieved tends to differ considerably from one institution to another and is indicative of a given institution's promotional strategy. Indeed, promotional styles may emphasize and reinforce some universities' core values and tradition, while others tend to focus on establishing a trustworthy relationship with the students by enhancing proximity. These aspects will be examined in more detail in the following sections of this chapter.

2. Data and methodology

This study is based on a quantitative and qualitative analysis of 60 Italian university websites.[1] All the data have been retrieved from *4*

1 The websites examined in this study are those of the following universities, which have been selected because they display an introductory section: Avogardo Piemonte Orientale, Bari, Basilicata, Bergamo, Bicocca Milano, Bocconi Milano, Bologna, Bolzano, Cà Foscari Venezia, Cagliari, Calabria, Camerino, Cassino, Catania, Cattolica, D'Annunzio Chieti-Pescara, Federico II Napoli, Firenze, Foggia, Genova, Iuav Venezia, Iulm Milano, Iusm Roma, L'Aquila, Luiss Roma, Lumsa Roma, Macerata, Marche, Messina, Milano, Modena-Reggio Emilia, Molise, Normale Pisa, L'Orientale Napoli, Padova, Palermo, Parma, Pavia, Perugia, Pisa, Politecnico Milano, Politecnico Torino, Roma 3, Salento, Salerno, Sant'Anna Pisa, Sassari, Siena, Sissa Trieste, Teramo, Torino, Trento, Trieste, Udine, Vita-Salute S. Raffaele, Sapienza Roma, Università Stranieri Perugia, Urbino, Valle D'Aosta, Verona.

International Colleges & Universities (from June to August 2012), an on line directory and search engine that categorizes and reviews international higher education institutions (see Caiazzo 2009, 2011).[2] The corpus is made up of 25,398 tokens and consists of texts found in the introductory sections (henceforth IS) of the various universities considered. I have classified the ISs as follows: *About Us*[3], *The university*, *University of* […], *Welcome to the university of* […], *The University in brief*, *Presentation of the University*, *Mission Statement*, *Relazioni internazionali*, including *Message from the Rector/Chancellor*. Each IS varies in its emphasis of the university's history, tradition, and future development.

Welcome Messages from the Dean were also included in the corpus where available. In at least five institutions' websites, the Dean's Message to the students stands alone and replaces the IS, while in the remaining ones a more traditional format has been adopted. Where the two are combined, the Dean's message accompanies the more formal and descriptive, albeit equally promotional, IS.

In most ISs, the English versions of the websites are either translated from the original Italian text or directly written in English to address international students' needs. The location and the icons of the English versions tend to vary. In 50 websites the icon is located either on the upper right or left hand corner of the homepage, under the heading 'English version', or is simply indicated by a UK flag icon. Access is otherwise possible via the UK flag and its accompanying text. In the remaining ten websites, neither a flag nor text is provided and the English version is accessible via a menu section under the heading *International students*. In the latter case, international students, or Erasmus exchange program students, are these universities' prime target audience.

---

2 The website is: <http://www.4icu.org/>.

3 I have observed that the heading *About Us* is sometimes substituted by *About* + the university's name or simply by *About*.

3. Analysis

The specific pronouns *we, you* and the possessive adjectives *our* and *your* appear in 60% of the websites, even though they are not all used at the same time, and tend to be used in varying degrees and ways. The data collection procedure consisted of downloading the ISs from the universities' websites, and processing the relevant data with *WordSmith Tools* 5.0 (Scott 2008) to create a word list. Secondly, with *WordSmith Concord Tool* (Scott 2008) concordances of the pronouns were obtained. In analyzing the co-text (Baker 2006; Hunston 2002) it was clear how specific pronouns determine whether the reader is included or excluded. Moreover, the verbs appearing near the pronouns *we* and *you* were analyzed on the basis of the categories established by Biber *et al* (1999). The same approach was taken regarding the possessive adjectives *our* and *your*, which usually collocate with a noun.

At this point it is worth recalling Biber *et al*'s verb categories which are broken down into the following domains. Firstly, Activity verbs refer to "actions and events that could be associated with choice, and so take a subject with the semantic role or agent" (Biber *et al* 1999: 361); some examples of Activity verbs are: *make, get, go, come, buy, play, produce,* and so on. Secondly, Mental verbs indicate activities and states typically attributable to humans, and include verbs such as: *know, think, want, hope, decide, discover, feel*, and so on. Thirdly, Existence or Relational verbs typically describe a state, a relationship, or a link that exists between two elements; examples of these verbs include: *be, seem, appear, have, contain, include, involve*, etc.

There are also Aspectual, Communication, Causation, and Occurrence verbs. Aspectual verbs, indicating actions, progression, or the state of progress of actions, and Communication verbs, indicating communication activities such as *speak, call, discuss*, have been included in the Activity verb domain due to their statistical insignificance. Lastly, Causation verbs describe actions entailing a change to a new situation, or state of affairs and refer to verbs such as:

allow, cause, enable, force, help, let, and Occurrence verbs which describe physical events that occur independently of people's actions; examples of these verbs include: *develop, become, occur*, etc. (Biber *et al* 1999: 361-364). In this corpus the most relevant verbs analyzed are Activity, Mental and Existence ones; Causation and Occurrence verbs rarely occur.

The following table shows the number of pronouns and possessive adjectives in the university websites being studied:

Pronouns	*Occurrences*
WE	75
YOU, both in subject and object positions	85[4]
OUR	123
YOUR	36

Table 1. Number of pronoun occurrences.

As apparent from the table, the possessive adjective *our* is by far the most common, followed by *you*, *we* and finally by the possessive adjective *your*. In the following sub-sections, data concerning each pronoun in the corpus are presented and discussed.

3.1. The roles of 'we' between exclusiveness and inclusiveness

As might be expected, the *we* that excludes the reader is found in the majority of instances in the ISs. As Fairclough has pointed out (1993: 146-147), the use of *we* is used to 'personalize' the university; moreover, this use of the exclusive *we* allows the institution to construct a more corporate identity (Benwell/Stokoe 2006: 118). This 'corporate' university identity includes the university's board and administration, along with its teaching and administrative staff, and contributes to portraying a cohesive, highly reputable, socially

4 For the purposes of this study only *you* in the subject position was analyzed.

responsible organization (Caiazzo 2011), as will be seen in the following examples.

There are 75 *we* forms in the corpus; 83% of them are exclusive of the reader and thus refer only to the university, while the remaining 17% have been categorized as inclusive of their addressee. Interestingly, notwithstanding the exclusive or inclusive role of pronouns they could still be regarded as multifunctional, as they *potentially* include or exclude their audiences. This ambivalence is pragmatically significant as it enhances the degree of their readers' involvement. As far as the verbs that collocate with an exclusive *we* are concerned, concordance lines generated with *WordSmith Tools* 5 have shown that Activity verbs appear in 39% of the cases. The following examples show instances of *we* + Activity verbs:

(1) […] *we* are striving to keep pace with the ever-faster changes of our society of the modern world […][5].

(2) *We* work hard to create a vibrant intellectual environment […].

(3) *We* are offering courses – Bachelors, Masters, PhDs - which will equip you for the challenges of a complex and changing world.

The main trend seems to be that universities are becoming both providers of resources (e.g. education, courses, campus facilities) and organizations that wish not only to preserve their existing reputation but also to extend and build on it. In particular, the use of the progressive verb form in (1) and (3) attaches a dynamic quality to these universities' profiles and reinforces their commitment to improving services. Interestingly, most of the *we* instances in the corpus are found in the Chancellors' messages because they address students in a more direct and personal way. These texts potentially create a closer rapport with the addressees than the ISs' slightly more objective messages do. Moreover, these 'personal' messages permit a more fluid description of the institution, especially when the

---

5 All the quoted English texts in this paper have been taken directly from the university websites. No changes were made to any of these texts by the author of this paper.

university is relatively new and therefore does not have a long-standing academic record:

(4) The university of [...] still has some problems (why hide them?), above all with regard to its buildings, but *we* are constantly working to resolve these problems.

In example (4), the Chancellor's message plays down the university's shortcomings, albeit without hiding them; the tone of the message allows him to strike a balance between the university's strengths and weaknesses and to emphasize how the university is totally committed to improving the quality of its premises and of its services.

Much like Activity verbs, Mental verbs occurring with *we* total 33% of the corpus. Significantly, most Mental verbs are connected to students and, in at least two instances, are potentially inclusive of them, and possibly impersonal, as shown in the examples below:

(5) [...] especially if *we* consider the history of Southern Italy and Calabria, which on few other occasions have achieved similar outstanding and meaningful results.

(6) Studying at the [...] means opening out to the knowledge of people from different linguistic, cultural, religious, and social backgrounds and learn to live together. In order to do this *we* need to get to know each other, and to go beyond traditional stereotypes which often create misunderstanding, if not hostility.

Significantly, in addition to the references to students, (5) and (6) also include other subjects by alluding to the local and national communities with which the university shares the same national heritage and cultural background. The university becomes the bearer of an important social, cultural and humanitarian message. This message has a twofold aim: on the one hand it encourages addressees to be open-minded and tolerant towards other people and world cultures, while on the other, it outlines the qualities that potential students should have. Moreover, the university's commitment to social and cultural issues is confirmed by its curricula, which feature the teaching of world cultures and languages. Sentences (7) and (8) below containing Mental verbs, are exclusive of the addressees:

(7) Dear student, *we* would like to take this opportunity to welcome you to the university of [...].

(8) *We* hope to have the privilege of assisting you and helping you [...].

In (7) and (8) the university welcomes students and tries to establish a rapport of trust. By using Mental verbs in instances (9) and (10), the institution seeks to encourage students to apply by offering them help with practical problems, delivering high quality education, and envisaging promising postgraduate or employment opportunities for them in the future:

(9) *We* accept the challenge of training competent, open-minded professionals [...].

(10) It is here, in this friendly and functional environment, that *we* see so many youths grow and train everyday, willing to work their way up to and export the University's reputation throughout the world.

Only 17 examples of Existence verbs emerge, making up 23% of the total *we* occurrences. Amongst these, nine instances collocate with *are/be* while only three collocate with *have*:

(11) [...] *we* are ranked at the top, in the Italian scientific scope [...].

(12) *We* are partners with more than sixty foreign universities [...].

(13) *We* therefore have an experienced, friendly, and well-qualified team of people [...].

(14) Now *we* have several main campuses in and on the edge of the city.

The university is deliberately depicted through the use of the Relational verb (have) as having special qualities and assets, such as extensive facilities and qualified personnel to ensure high-quality education.

As can be seen from the previous examples, the use of Mental verbs allows the university to establish and reinforce its reputation by projecting a positive and stable image; this image somehow counterbalances the dynamic and constructive, albeit somewhat slightly unsteady, picture that Activity verbs tend to convey about the university. This more solid and cohesive organization, achieved mainly through the use of *we* with Mental verbs, seems in line with what Luisa Caiazzo (2011) called an 'enticing *we*' in which the main aspects of the university (services, objectives, reputation) seem to work together harmoniously. According to Caiazzo (2011: 255), all these constructive elements described on their university websites are "intertwined with one another in a synergistic endeavour, thus contributing to construe a 'corporate' identity for the university in terms of what might be defined as an 'enticing *we*'. Whoever is not yet a member of the community has just to join".

3.2. 'You': constructing an academic rapport between the university and the student

In the corpus we found 85 instances of *you*. All of them refer to students, at times in more general terms and, at other times, in more specific terms. After counting and selecting all the sentences, 47 instances of *you* were found in the subject position (52%), while 38 of them (48%) are in the object one. As regards the verb forms occurring with *you*, 45% are Mental verbs and thus they appear more frequently than with the *we* forms. As in the case of *we*, the main function of Mental verbs is that of establishing a rapport with the student:

(15) [...] *you* remember our University as a place where *you* laid the foundation for your dreams and found the strength to make your life plans come true.

(16) This Guide is intended to help *you* get settled in and it provides practical information that we hope *you* will find useful.

(17) We are glad that *you* have chosen our University for your studies and we hope that *you* will find this experience rewarding in terms of education and personal development.

Whenever both Mental and Activity verbs are employed, as demonstrated in (15) through to (17) the use of *you* conveys a sense of familiarity with the students/addressees and indicates the addressers' willingness to create a bond with them. The institution presents itself as supportive of students on a practical and psychological level. At the same time, the institution thanks students for their choice and expresses the wish that they will fulfill their dreams and expectations.

In (18) the students are directly addressed with a question containing Mental verbs:

(18) Are *you* planning to spend a period of time abroad, for study or research purposes? Are *you* unsure which University will best serve your interests and best meet your professional/personal goals and expectations? International students have many excellent opportunities available to them […].

The use of questions in (18) strongly resembles an advertising style in which the writer tries to capture and engage the interest of the reader (Leech 1966). The questions are meant to prompt the interest of prospective students who intend to apply to an international study program, but at the same time, the questions appear rather provocative as the students are questioned about their broader goals in life. It is as if the addresser were asking the student: 'Do you have a plan or not? Are you ready to start a new life as an international student by applying to an international study program?' In other words, the two questions raise concerns about whether such prospective applicants consider themselves to be international students, or to even be students at all.

The rate of Activity verbs in sentences with the pronoun *you* represents 34% of the corpus. In the following examples with Activity verbs, the student is asked to support the university by performing specific actions; for example, giving suggestions, making proposals, and speaking with pride about the institution, thus actively contributing to building and enhancing the university's reputation:

(19) We would be grateful if *you* could provide any suggestions, comments, proposals to the International Relations Office.

(20) I hope that one day *you* can speak with pride about having been students at the Università degli Studi […].

In short, on the one hand, the institution presents itself as supportive of the students on a practical and psychological level, as the previous examples confirm; on the other, the university also invites support from the students to enhance its reputation and consequently to reinforce its students' commitment, as is apparent from (19) and (20); this is particularly evident in the examples with Activity verbs.

Existence verbs with the *you* pronoun represent 21% of the corpus. In the following examples of Existence/Relational verbs, the institution, as well as establishing a connection with prospective students and providing support for them, also endeavors to set them goals:

(21) [...] I hope *you* will have a stimulating experience of university work and life.

(22) My wish is that *you* will be an active member of our community and develop the spirit of affiliation and pride in being part of the University [...].

These examples, drawing on various uses of *you*, reveal the potential of this pronoun to involve the addressees and to establish a rapport with them. In this sense, the analysis of the use of *you* in this corpus does not differ substantially from the use of *you* in advertising and in promotional language generally. Moreover, while the addresser, *we*, is identified as the university, *you* also functions as a high-involvement strategy in trying to win the readers over by addressing them directly. It is almost as if the two pronouns *we* and *you* were participating in a dialogue and negotiating important aspects of university life and of the students' future. In the following sentence, *you* is in juxtaposition with *we* as *we* explicitly refers to the services the institution delivers to its students/customers:

(23) *We* will be at your side to help *you* build up your future!

A further question for consideration is the use of *you* as it is employed in both singular and plural references in English. This undoubtedly contributes to the doubly *exophoric* way of addressing a given audience (Lombardo 1999: 114), which is largely employed in promotional language. On one hand, *you* may refer to the student who is actually reading the website; on the other it may also represent students as a general target category. In this second case, *you* may represent the ideal prospective student who is deliberately constructed by the addresser. The pronoun *you* could almost pass as a 'ready-made product', in other words, someone who has good academic credentials and a desirable lifestyle. A prospective international student could easily identify with this profile and decide to apply to the university.

3.3. 'Our' in the ISs of university websites

As shown in Table 1, the possessive adjective *our* has 123 occurrences and is therefore the most frequent adjective across the corpus. It is used more frequently because of the importance universities place on their traditions and reputation; it reinforces the notion of strong and reliable institutions created and supported by people who, as in the case of *we*, strive to give them credibility. Just like the pronoun *we*, it is also possible to determine exclusive and inclusive uses of the possessive adjective *our*. The exclusive use of *our* is more common, as there are 108 occurrences in the texts (88% of all occurrences). This is consistent with the main purpose of these pronouns and adjectives' exclusive mode, which is, namely, to establish the institution's identity and describe its main features.

It is worth noting that 40 instances of *our* exclusive of the addressees collocate with *university*, and are often followed by an Existence/Relational verb:

(24) […] *our* university was built around the concept of the Campus and was, and still is conceived as a large laboratory of ideas and inspirations […].

(25) […] *our* university is a public teaching and research institution offering a full range of academic programmes […].

(26) *Our* university has around 30,000 students, with 50% from outside the region and about 15% who are foreign - the highest percentage in Italy.

These examples seek to project an image whereby the university's identity is defined by how it has been constructed and construed (24), by what it offers (25), and by reinforcing this image with relevant statistics, with a promotional aim (26). In (26) this description relies on quantification discourse and the use of a superlative to emphasize internationalization (Benwell/Stokoe 2006: 118). The Existence/ Relational verb forms used in these examples contribute to conveying a sense of stability along with an image of an almost autonomous, personified entity. Importantly, 85 occurrences of exclusive *our* collocate with nouns pertaining to the university domain, such as: *student*/s (11), *research* (4), *institution* (5), *mission* (3), *community* (2), *graduates* (2), *degrees* (1), *curricula* (1):

(27) […] *our* mission has remained unchanged to facilitate intensive academic achievement at the forefront of world-class education.

(28) *Our* institution is one of the oldest universities in Europe and the oldest one in Lombardy.

(29) Come to *our* university and share with us visions and analyses, projects and experiments: the challenge issued to all of us by the contemporary world is too complex to be faced on one's own.

The matters addressed in the examples above are intended to portray an image of a university which emphasizes tradition, history and high-quality teaching along with connections with the wider world.

In (29) the use of the imperative verb form ("Come to our university") strongly encourages students to enroll in a university and invites them to participate in a center of learning, underpinned by shared values and projects, which can help them prepare for challenges posed by contemporary society. The aim of this strategy is to make students feel part of the learning environment and to convince them that their studies will enable them to deal with the challenges of an increasingly demanding world.

Even though there are fewer instances of inclusive *our*, there is an important distinction to be made about its use. Five out of 15 of these occurrences with *our* (4%) are inclusive of the reader, meaning that both writer and reader are supposed to share the same actions or cognitive states. However, there are ten instances (8%) in which *our* includes local students, the local/national community, but it does not necessarily include every reader/student. It is somewhat indirect and conveys potentiality rather than inclusiveness. In other words, not every reader might feel included, whereas other readers may take up this kind of invitation. As within inclusive *we*, then, a decision was made to consider 'potential inclusiveness' as a significant category; this rather appropriately reflects what Benwell and Stokoe (2006: 118) observe: "The ambiguity of the target addressee is suggested in the shifts between positioning students as *part* of the enterprise by the use of the possessive 'our' [...], and positioning them as external but potential". The following examples confirm this tendency :

(30) [...] to join in with the dynamic and sustainable positioning of *our* region within the national economic and social context.

(31) [...] the cultural contributions and innovations that had been experimented for some time both in Europe and in the USA, were imported to *our* country and transferred to the national university system.

(32) This, combined with the many attractions of *our* very special city.

What clearly emerges from these statements is that not all addressees will feel as though they belong to the university, while others will feel as though they do. Therefore, this potential inclusiveness may appeal to Italian students and the local community reasonably well since it projects an image of a university which is very much connected to the local community's social fabric. This interpretation reveals how the institution is attempting to connect itself with the rest of the country and to emphasize this relationship to prospective foreign applicants.

The examples of inclusiveness are similar to those regarding the inclusive *we* in which general comments were made concerning society at large. Similarly, these instances vary in their degree of inclusiveness of the addressees:

(33) *Our* aim is to encourage *our* students' creativity, enhancing their capacity to engage with the great challenges of *our* society.

(34) [...] *our* so called "Advanced Society", runs the risk of becoming a society where human exchanges are reduced to an Internet computer screen.

(35) Therefore, we need *our* cultural and social contacts with other people and countries to proceed through mutual opportunities […].

In (33) the first two instances of exclusive *our* refer to the university, whereas the third instance of *our* includes everyone, despite the fact that 'society' could be interpreted as having different degrees of inclusiveness, for example a 'national society', or even a 'Western one'. Example (34) has an international flavor; the notion of an 'advanced society' clearly transcends national boundaries, while the use of *our* is intended to include everyone. The same could be said for (35) in which *our* may refer to the people from the same nation, but potentially it could also include everyone. In these examples, when compared to *we*, *our* suggests a stronger sense of sharing at a personal level, which is what the university evidently wants to convey through the ISs.

3.4. *The use of 'your' in constructing a life of a potential student*

As one might expect, most of the 36 instances of the possessive adjective *your* are inclusive of the students. The use of *your* in the IS aims to construct the image of the 'ideal' student so that the potential student can identify with him/her and thereby take on the same identity. Thus, 17 instances of *your* collocate with nouns pertaining to the university world, as can be seen from these examples: *study/studies* (5), *university journey* (1), *university career* (1), *academic studies* (1), *training* (1), *lab* (1) *class* (1), *professors* (1), *courses* (1):

(36) […] you can walk from *your* class to your Residence, from *your* lab to the canteen.

(37) In this user-friendly site, you will find links to *your* professors and the administrative offices.

(38) […] make the best of *your* courses from the very beginning.

A clear effort is being made to depict student life as appealing, as well as presenting potential incoming foreign students with a realistic picture of their duties as students: a student has to study independently, choose courses, go to class and to the lab, meet professors. An emphasis is also placed on broadening the scope of campus life activities. The rest of the nouns collocating with *your* describe more general aspects of a student's life on campus, whilst still being connected to the university; these include: *(your) time* (4), *first time* (1), *future* (2), *stay* (2), *experiences* (1), *dreams* (1), *life plans* (1).

In addition to promoting its facilities and orientating students toward their studies in a rather matter-of-fact fashion, the university also wants to appear supportive on an emotional level:

(39) We will do our best to help you enjoy and benefit from *your* experiences whilst here.

(40) [...] we try to make *your* time as enjoyable and successful as possible.

(41) [...] you remember our University as a place where you laid the foundation for *your* dreams and found the strength to make *your* life plans come true.

There is a tendency here to construct an identity for the student which goes beyond the purely academic sphere. This is achieved by referring to aspects of university campus and students' social life, by encouraging students to look beyond the physical and cultural boundaries of the university itself, and by encouraging them to envisage their future.

4. Concluding remarks

This chapter has shown that, far from being a random and casual choice, the use of personal pronouns in the texts analyzed follows a promotional strategy which closely mirrors corporate strategy, where the universities' ultimate goal is, obviously, to convince students to enroll in the university. Moreover, to that end, the university creates a bond with the student by establishing an apparently friendly relationship; the university goes on to assume the role of a reliable advisor, almost a virtual online coach, intended to benefit the student's future cultural and professional life.

By analyzing the use of the pronouns *we* and *you*, and the possessive adjectives *our* and *your* we have seen how universities and prospective students are the protagonists of the ISs of university. The university, in particular, is often personified and thus becomes the provider of a number of assets, which become advantageous for both the university and the students. Most universities construct their identity upon a solid tradition, a good reputation, and well-established ties with the surrounding region and city, while at the same time constructing an ideal student life for future international students. They do this by underlining the high-quality of their courses, the competence and experience of their academic staff, their innovative lab facilities and their exciting and rewarding campus life. Such knowledge and infrastructure provide an ideal springboard for promoting the personal growth and development of foreign students studying in Italy.

This study has also shed light on an important aspect concerning the English version of ISs of Italian university websites. In addition to showing the various ways in which promotional tools such as personal pronouns and possessive adjectives are used (however it is worth remembering that 40% of the websites do not use personal pronouns or adjectives in their ISs), this chapter's findings indicate that a number of highly diverse texts emerge from these university websites. These range from rather traditional and dryly informative ISs, which do not use pronouns or possessive adjectives to communicate with their target audience, to more innovative, inclusive and creatively dynamic texts. This approach, which has much in common with corporate advertising, has been deployed by a widespread use of pronouns and possessive adjectives and appears to have been adopted to engage with a wider audience and to enhance the universities' marketing appeal. Further research on the evaluation of the pragmatic power of these traditional and innovative texts could contribute to determining their impact and effectiveness while encouraging Italian universities to adopt common guidelines geared towards a more coherent and cohesive style of self-presentation.

References

Altbach, Philip G. 2004. Globalization and the University: Myths and Realities in an Unequal World. *Tertiary Education and Management* 10, 3/25.

Askehave, Inger 2007. The Impact of Marketization of Higher Education Genres – The International Student Prospectus as a Case in Point. *Discourse Studies* 9/6, 723-742.

Baker, Paul 2006. *Using Corpora in Discourse Analysis*. London: Continuum.

Benwell, Bethan / Stokoe, Elizabeth 2006. *Discourse and Identity*. Edinburgh: Edinburgh University Press.

Biber, Douglas / Johansson, Stig / Leech, Geoffrey / Conrad, Susan / Finegan, Edward 1999. *Longman Grammar of Spoken and Written English*. London: Pearson Education Limited.

Caiazzo, Luisa 2009. 'National', 'international' and 'global' across British and Indian university websites. *ESP Across Cultures* 6, 45-59.

Caiazzo, Luisa 2010. The Promotional English(es) of University Websites. In Cagliero, Roberto / Jenkins, Jennifer (eds) *Genres, Discourse Communities and Global English*, Bern: Peter Lang, 43-59.

Caiazzo, Luisa 2011. Hybridization in Institutional Language: Exploring *we* in the 'About us' Page of University Websites. In Sarangi, Srikant / Polese, Vanda / Caliendo, Giuditta (eds) *Genres on the Move. Hybridization and Discourse Change in Specialized Communication.* Napoli: Edizioni Scientifiche Italiane, 243-260.

Cheney, George 1983. The Rhetoric of Identification and the Study of Organizational Communication. *Quarterly Journal of Speech* 69/2, 143-158.

Crawford Camiciottoli, Belinda 2004. Walking on Unfamiliar Ground: Interactive Discourse Markers in Guest Lectures. In Partington, Alan / Morley, John / Haarman, Louann (eds) *Corpora and Discourse*. Bern: Peter Lang, 91-104.

Crawford Camiciottoli, Belinda 2005. Adjusting a Business Lecture for an International Audience: A Case Study. *English for Specific Purposes* 24, 183-199.

Dewey, Martin 2007. English as a Lingua Franca and Globalization: An Interconnected Perspective. *International Journal of Applied Linguistics* 17/3, 332-354.

Fairclough, Norman 1993. Critical Discourse Analysis and the Marketization of Public Discourse: The Universities. *Discourse & Society* 4/2, 133-168.

Fairclough, Norman 2006. *Language and Globalization*. London/New York: Routledge.

Ferrarotti, Laura 2009. The Newsreader and the TV Audience: A Comparative Perspective of the Use of 'we' and 'you', in Haarman, Louann / Lombardo, Linda (eds) *Evaluation and Stance in War News. A Linguistic Analysis of American, British and Italian Television News Reporting of the 2003 Iraqi War.* London: Continuum International Publishing Group Ltd., 72-96.

Fortanet, Inmaculada 2004. The Use of "we" in University Lectures: Reference and Function. *English for Specific Purposes* 23, 45-55.

Hunston, Susan 2002. *Corpora in Applied Linguistics*. Cambridge: Cambridge University Press.

Hyland, Ken 1998. Exploring Corporate Rhetoric: Metadiscourse in the CEO's Letter. *Journal of Business Communication* 35/2, 224-244.

Hyland, Ken 2002. Directives: Argument and Engagement in Academic Writing. *Applied Linguistics* 22/3, 215-239.

Kuong, Julia 2000. Introduction: Marketization and Privatization in Education. *International Journal of Educational Development* 20, 87-92.

Leech, Geoffry 1966. *English in Advertising: A Linguistic Study of Advertising in Great Britain*. London: Longman.

Lombardo, Linda 1999. Advertising as Motivated Discourse. In Lombardo, Linda / Haarman, Louann / Morley, John / Taylor, Christopher (eds) *Massed Medias. Linguistic Tools for Interpreting Media Discourse*. Milano: Led.

Medina Bolaños, Alicia / Medina Rodríguez, Maria Jesus / Medina Bolaños, Lydia / García Losada, Luis (eds) 2005. Analysing Digital Genres: Function and Functionality in Corporate Websites of Computer Hardware. *Ibérica* 9, 125/147.

Osman, Hajubah 2008. Re-branding Academic Institutions with Corporate Advertising: A Genre Perspective. *Discourse & Communication* 2, 57-77.

Rounds, Patricia L. 1987. Multifunctional Personal Pronoun Use in Educational Settings. *English for Specific Purposes* 6/1, 13-29.

Salvi, Rita / Bamford, Julia (eds) 2007. *Business Discourse: Language at Work*. Roma: Aracne.

Scott, Michael 2008. *WordSmith Tools*, version 5.0. Oxford: Oxford University Press.

Swales, John M. / Rogers, Priscilla S. 1995. Discourse and the Projection of Corporate Culture: The Mission Statement. *Discourse & Society* 6/2, 223-242.

Walsh, Polly 2004. A Complex Interplay of Voices: First and Second Pronouns in University Lectures. In Anderson, Laurie / Bamford, Julia (eds) *Evaluation in Oral and Written Academic Discourse*. Roma: Officina, 31-52.

Wernick, Andrew 1991. *Promotional Culture. Advertising, Ideology and Symbolic Expression*. London: Sage Publications.

GIUSEPPE BALIRANO / MARIA CRISTINA NISCO

Identity and the Linguistic Construction of *Space* and *Place* in Diasporic Films: the *DesiCorpus*[1]

1. Introduction

Identities have often been defined in linguistic studies as inconsistent, constructed, dynamic, changing, and hybrid entities or processes (Le Page/Tabouret-Keller 1985; Cameron 2000; Hall 2006; Byrd Clark 2009) which tend to be dynamically negotiated as discourse unfolds (De Fina/Schiffrin/Bamberg 2006). According to Hall (2006), the personal construction of cultural identities is progressively becoming more problematic since it pulls in multiple directions; discourse participants consciously shift from one identity to another, becoming multiple people in multiple places, and sometimes performing various identities (Byrd Clark 2007, 2009; Balirano/Guzzo 2011). De Fina (2003: 15) emphasizes the complexity of identity because language users often draw upon several identities in a given communicative context. Moreover, it is acknowledged that since individuals belong to varied groups, they take on a variety of identities established by their memberships in these groups. These identities can be seen as tied to social practice and interaction and to processes of differentiation from

1 This chapter has been written in the context of the Italian national research project (PRIN 2008); the local research unit, based at the University of Naples L'Orientale and coordinated by Professor Julia Bamford, focused on "The use of space in the discursive construction and display of cultural and professional identities". The analysis uses the tools of corpus analysis, looking in particular at how space is represented discursively and how this is related to the construction of both cultural and disciplinary identity. The authors discussed and conceived the article together. In particular, Giuseppe Balirano is responsible for Sections 1 and 2; Maria Cristina Nisco is responsible for Sections 3, 4 and 5.

other groups. An emerging focus on language and identity studies pertains to the creative formation of hybrid identities through speech stylisation and language crossing. This rising attention is possibly due to global migration, which has brought individuals and groups from different countries into direct contact with each other. According to Rampton (2009: 149) discourse participants produce specially marked representations of languages, dialects, and personal styles that go beyond their own usual linguistic repertoire: "Crossing […] involves a stronger sense of social or ethnic boundary transgression". Rampton's main interest was the investigation of the ways in which mixed-race youths in Britain used language to construct hybrid identities. The groups were of mixed ethnic origins including London teenagers together with youths of Caribbean, Indian and Pakistani descent.

A significant contribution to the analysis of identity construction by means of language is offered today by the mainstream film and media industry which seems to be increasingly involved in the difficult process of identity representation of first and/or second-generation immigrants in Western countries. A variety of partly overlapping terms has been conceived to refer to films regarding diasporic subjects in general, among them 'intercultural cinema', 'transnational cinema', 'hyphenated identity cinema', 'accented cinema' (Naficy 2001), but a specialised genre connected to the South Asian diaspora has recently gained recognition and popularity: Desi films.

Desi (or Deshi) is a Sanskrit word literally meaning 'from India' or 'of the country', which is generally used to describe a person of South Asian origin (with ancestry from India, Pakistan, Bangladesh or Sri Lanka) living outside of South Asia. While tackling the tricky questions of identity, belonging and home, Desi films seem to articulate a new diasporic narrative representing the South Asian diaspora through multimodal texts in which the discursive construction of identities is often linguistically indexed through labels, code choices, lexical choices, stances, styles or linguistic structures (Bucholz/Hall 2005). In such hybrid linguistic and cultural contexts, new forms of self-representations may emerge as important discursive practices favouring a predictable linguistic expansion and renewal, while challenging dominant power structures:

> Recent Desi re-presentations seem to facilitate the circulation of a new diasporic narrative, one which construes the unresolved dilemmas of 'identities' and 'home'. Thus, contemporary diasporic self-representations can no longer be considered a simple imitation or reflection of the mainstream, but real forms of identity construction; they are discursive practices which are carried out by diasporic subjects within specific power relationships. A new stability is thus enforced by a hybrid optic which offers an alternative and visible practice, interconnecting and negotiating identities by means of different media and idioms; thereby producing original forms of self-identification with the homeland. (Balirano 2009: 87)

The encounter/clash between English and non-English semiotic and more general linguistic systems requires forms of 're-contextualisation' and creative mixed language use generating linguistic creativity which enacts new balances between discourses and audiences. Therefore, the audiovisual translation of diasporic filmic productions necessarily calls for new strategies to render acts of identity, markers of resistance, abrogation, appropriation in order to re-place (and/or re-locate), and 're-root' the English language(s) in the target one.

Since the major diasporic communities of South Asian origins are concentrated in the United Kingdom and in the United States – where they have made significant contributions in several fields that are of interest to this research (ranging from the media to culture and language) – self-representations in filmic productions of such communities were chosen as the focus of this chapter.

2. Identity and the discursive use of space/place

A number of linguistic studies have investigated forms of identity representation in various discourse domains, communities and the media. In particular, previous research shows how identity cues are revealed through some specific linguistic features, such as the use of first and second pronouns and specialized lexis. First and second person pronouns may also play an important role in constructing social identities because they can immediately provide useful, overt or unconscious, information about the interpersonal relations of the discourse participants. Moreover, in the linguistic investigation of space/place as the location of discourse, indexicality also plays a key role in anchoring language to speakers in specific contexts and in signalling acts of identity construction.

The connection between space and the discursive construction of identity in Desi films seems to develop around two main ideas: place/space as produced *in* and *as* a topic of discourse, and place/space as the location for discourse (Benwell/Stokoe 2002, 2007). Identity practices can be closely linked to the linguistic use of space, in particular with regard to ethnic identities, thus showing that this is not only tied to language but also to other semiotic systems – such as media productions – and practices in interaction. Space and place can be seen either as location or mental spaces related to the public and the private, to memory and identity (Manzo 2003). By using for example locatives, personal pronouns and time expressions, language users point to their roles not only as discourse producers and recipients, but also to their location in time and space and to their relationship to others (De Fina/Schiffrin/Bamberg 2006). This chapter will adopt a micro approach to see how Desi hybrid identity is construed linguistically through a corpus-driven investigation of transcribed spoken discourse and its comparable translation in Italian.

3. Corpus and methodology

An initial selection of English-language Desi films available in their Italian versions was made by browsing the Web, especially surfing the website *Migrant and Diasporic Cinema in Contemporary Europe* (henceforth MDCCE),[2] which provides a useful database and detailed information on the topic. With the aim of exploring the evolution of

2 <http://www.migrantcinema.net/> (unless otherwise specified, all websites were last accessed in October 2012).

contemporary Europe over the past twenty-five years and the ways in which multi-cultural and multi-ethnic presences and themes address a shift from national to transnational identities, the website offers a specific section on 'British Asian Cinema'. Further research was then carried out by surfing the *Black British Film Screenonline*[3] which also hosts an 'Asian-British Cinema' section with films focusing on the South Asian experience in the UK.

For all the films in the database, the MDCCE website features a link to the *Internet Movie Database* that also gives the Italian titles, whenever a dubbed version of the film has been produced (while subtitled films were not included in the results). At this stage of the search, 21 English-language films thematically connected to the South-Asian diaspora in Great Britain or the US could be retrieved with the corresponding dubbed versions in Italian, as shown in Table 1.

Original (English)	***Dubbed (Italian)***
Heat and Dust (1982, James Ivory)	*Calore e polvere*
My Beautiful Laundrette (1985, Stephen Frears)	*My Beautiful Laundrette – Lavanderia a gettone*
Sammy and Rosie Get Laid (1987, Stephen Frears)	*Sammy e Rosie vanno a letto*
Madame Sousatzka (1988, John Sclesinger)	*Madame Sousatzka*
Salaam Bombay! (1988, Mira Nair)	*Salaam Bombay!*
Mississipi Masala (1992, Mira Nair)	*Mississipi Masala*
Bhaji on the Beach (1993, Gurinder Chadha)	*Picnic alla spiaggia*
Brothers in Trouble (1995, Udayan Prasad)	*Fratelli nei guai*

3 <http://www.screenonline.org.uk/film/id/475617/>. The BFI website, which describes itself as a guide to Britain's film and TV history, also highlights that the British film industry has begun to wake up to the 'brown pound' since British-Asian films regularly enter the British top ten box-office charts, moving from margin to mainstream, and winning considerable public and critical acclaim. Whereas many early British-Asian films in the 1970s and 1980s were drama-documentaries, in more recent times – and especially since *Bhaji on the Beach* – British-Asian cinema has acquired increasingly mainstream features, and a distinctively South Asian diasporic cinematic practice has also emerged (Malik 2002).

My Son the Fanatic (1997, Udayan Prasad)	*Mio figlio il fanatico*
East is East (1999, Damien O' Donnell)	*East is East*
Everybody Says I'm Fine! (2001, Rahul Bose)	*I segreti si scoprono dal parrucchiere*
Monsoon Wedding (2001, Mira Nair)	*Matrimonio indiano*
Bend It like Beckham (2002, Gurinder Chadha)	*Sognando Beckham*
In This World (2002, Michael Winter-bottom)	*Cose di questo mondo*
Ae Fond Kiss (2004, Ken Loach)	*Un bacio appassionato*
Bride and Prejudice (2004, Gurinder Chadha)	*Matrimoni e pregiudizi*
Touch of Pink (2004, Ian Iqbal Rashid)	*Scandalo a Londra*
The Mistress of Spices (2005, Paul Mayeda Berges)	*La maga delle spezie*
My Bollywood Bride (2006, Rajeev Virani)	*Se scappi ti trovo*
The Namesake (2006, Mira Nair)	*Il destino nel nome*
Brick Lane (2007, Sarah Gabron)	*Brick Lane*

Table 1. Full list of English-language Desi films in the original and dubbed languages.

However, due to the large amount of data to be processed, we further narrowed down our search field and criteria to make our corpus of Desi films, which we then named *DesiCorpus*, more manageable while still being representative. To this purpose, we have exclusively focused on the films produced by diasporic subjects, in a time-span ranging from the year 2000 to 2007. Therefore, since audio-visual translation research mainly relies on a case-study approach (Gambier/Gottlieb 2001), this chapter concentrates on samples from a corpus of 11 films in their original language(s) and their dubbed versions. The corpus amounts to a total running time of approximately 35 hours and 20 minutes of transcribed audiovisual material (equivalent to about 115,000 words), comprising both the English versions and the Italian dubbed films, and constituting two parallel, bilingual[4] sub-

4 Or rather multilingual, in the sense that some original versions display several code-mixing/switching between English and various other South Asian languages, such as Punjabi, Urdu, Hindi, and so on.

corpora of dialogue-based language that is meant to reproduce spontaneous speech and naturally-occurring face-to-face conversations.[5] In addition to the textual corpus, a further database will be created in order to gather the multimedia material and show the video shot of individual scenes in the two different language versions of the same film.[6]

Once the scripts of all the films were available in computer-readable files,[7] we proceeded with the linguistic analysis. Special attention was paid to particular linguistic features to identify the changes and similarities emerging in the translational process from English to Italian as far as dubbing is concerned. Among the many markers of identity discourse, such as personal pronouns, culture-bound words, specialised lexis, code-mixing, code-switching, phonetic variation, special functions (intentional or not of ethnicity and hybrid identity), focus on deictic forms in English and Italian was the starting point of our analysis, due to the fact that specific references to geographical, cultural and temporal settings are the carriers of identity values that sometimes undergo a series of changes when they are rendered from one language to another. Therefore, in order to study the linguistic choices adopted both in the original and dubbed versions of the films to represent the discursive process of diasporic identity-building, qualitative observation was carried out to isolate instantiations of space and place within the scripts of the films, along with

---

5 Whether or not film dialogue can be considered spoken language is a debated question since it is produced under particular – hybrid – conditions: it is to be regarded as a form of 'written to be spoken as if not written' language (see Gregory/Carroll 1978; on the issue of natural speech as opposed to speech on screen see also Rossi 2002).

6 The textual and audiovisual corpus Forlixt 1 (Forlì Corpus of Screen Translation, 2003) together with the Pavia Corpus of Film Dialogue (2006) have already significantly responded to the need for more empirical studies in the field of audiovisual translation.

7 Transcribing films is often regarded as a thorny process that always implies an act of interpretation since there are no standardised conventions regulating the transcription of filmic oral data (Lapadat/Lindsay 1999). The films in the *DesiCorpus* were transcribed with simple orthographic transcriptions to which further information concerning linguistic variation (non-standard pronunciation, different accents) was added in order to have a record of as many linguistic phenomena as possible (see the prosodic transcription proposed by Freddi/Pavesi 2009).

quantitative analysis that determined the frequency and occurrences of a series of relevant linguistic items within the *DesiCorpus*. Such an investigation shed light on some specific linguistic and translational strategies which highlight the potential cultural power emerging from the textual positioning of diasporic subjects. Apart from contextualising the emerging data, this chapter aims to verify whether some specific linguistic patterns related to certain cultural elements do occur and what values are attached to them. Although the results of the analysis we are about to discuss are still at an initial stage, they are representative of the preliminary phase of our research (later steps – that are, however, beyond the scope of this chapter – will also concentrate on non-verbal, audio-visual aspects requiring a multimodal analysis of the *DesiCorpus*).

The choice to concentrate on deixis depends on one of its main features: it is extremely common in conversation, since in everyday language we heavily rely on deictic elements to refer to people, objects, places, and times, which makes them particularly interesting for linguistic analysis. A frequently quoted definition of deixis describes it as "the location and identification of persons, objects, events, processes and activities being talked about, or referred to, in relation to the spatio-temporal context created by and sustained by the act of utterance" (Lyons 1975: 63). In this view, deixis would be the anchorage of an utterance in the extralinguistic context in which it occurs (Fillmore 1997), it could be conceptualised as "a mental space in which the speaker and the addressee are co-present at a given moment in time" (Marmaridou 2000: 100).

A quantitative investigation of our *DesiCorpus* has shown that filmic discourse participants of different ethnic origins very often rely on a frequent use of spatial deixis in order to position and/or distance themselves from their mother-country or from their 'new' host country. Indeed, according to Thomas, spatial deixis "do[es] not mean much in isolation, it is only when you know where the speaker is standing or what the speaker is indicating that [it] become[s] truly meaningful" (Thomas 1995: 9). With the help of *Wordsmith Tools 5.0* (Scott 2008), we looked at how space is discursively characterized in our parallel corpora, and in particular how this relates to the construction of the speakers' positioning in English and in the Italian dubbed versions. Table 2 indicates the occurrences of spatial deixis in both sub-corpora:

Deictic expressions in EN	***Occurrences***	***Deictic expressions in IT***	***Occurrences***
HERE	85	QUI/QUA	60
THERE (local adverb)	72	LÌ/LÀ	50
US	34	STATI UNITI/ AMERICA	20
ENGLAND/UK	43	INGHILTERRA/ GRAN BRETAGNA	26
INDIA	67	INDIA	58
ABROAD	34	ALL'ESTERO	20
HOME	44	CASA/PATRIA	31
GO*	94	ANDARE	80
COME*	98	VENIRE	85
LEAVE*	34	PARTIRE/LASCIARE	21

Table 2. Full list of deictic expressions in the *DesiCorpus*.

As we can infer from the afore-mentioned occurrences, there are interesting irregularities in the two versions of our *DesiCorpus*. It partly depends on the fact that in the audiovisual translational process a reduction of original information is to be expected, but the Italian dubbed version is reduced to the point of not enabling us to easily identify instances of identity enacted by spatial deictic expressions.

Due to the complexity of the subject and the limited space available, among the different types of deixis – space, time, person – we exclusively took into account instances of spatial deixis,[8] providing selected and representative examples of descriptive and contrastive analysis found in some of the films of the corpus, and in particular: *Bride and Prejudice* (henceforth BP) and *The Namesake* (TN).

8 Spatial deixis comprises local adverbs (like *here* and *there*), demonstratives, deictic verbs (movement verbs like *to go* or *to come*), denoting spatial location in relation to the location of the speaker, following the fundamental distinction proximity-distality, 'close to the speaker' vs. 'far from the speaker' (see Anderson/Keenan 1985: 259).

4. Indexicality: changing spaces, changing identities

4.1. Bride and Prejudice

Bride and Prejudice, a 2004 musical film directed by Gurinder Chadha, is a Bollywood-style remaking of Jane Austin's novel *Pride and Prejudice* (1813). It is the story of Lalita Bakshi who meets Will Darcy, a handsome and wealthy American businessman she finds very arrogant and conceited, and with whom she develops a love/hate relationship. Chadha follows Austen's plot, but transplants it to modern India. In the first scene under investigation, Lalita and her parents are having dinner with Kholi Saab, a rich, clumsy and ostentatious Americanised relative who will later propose to Lalita.

Source text	***Target text***
Kholi: You must join me in US, you must, you must, you must. Eh?! I could help you start a business there. That's where the money's to be made. UK's finished, India's too corrupt.	Kholi: Mi deve venire a trovare in America. Promesso, promesso, promesso, eh?! Potrei darle una mano a mettere su un business. E' il posto ideale per fare soldi. L'Inghilterra ormai è finita e l'India è troppo corrotta.
Mr Bakshi: Kholi Saab, India is still a young country since independence. I hardly think its potential is over.	Mr Bakshi: L'India è ancora un giovane paese da poco indipendente. Il suo potenziale non credo sia esaurito.
Lalita: What do you think your US was like after 60 years from independence? They were killing each other with slavery and blindly searching for gold.	Lalita: Come pensi che fossero gli Stati Uniti dopo 60 anni di indipendenza? Non facevano altro che ammazzarsi per gli schiavi e per trovare l'oro.
K.: I'll tell you something, I've made a fortune as an accountant in LA. You know, the Indian community there is very professional, all doctors and computers, and not like these uneducated minicab/7-11 store types. It's ours for the taking. Anyone can become an American.	K.: Sa che le dico? Si guadagna molto facendo l'avvocato a Los Angeles. La comunità indiana là è molto professionale. Tutti dottori, con i computer, non come questi tassisti incivili e venditori ambulanti. Dipende da noi. Chiunque può diventare americano.

Table 3. Geographical deixis in *BP*.

Table 3 shows that, in both versions, spatial deixis (instances in the texts are in bold) is achieved through a variety of toponyms repeatedly mentioned by the characters involved in the conversation (US, UK, India, LA), thereby precisely locating each of them in a specific space – Kholi's space is America, whereas Lalita and her father's space is India. While boasting his 'new American roots', Kholi continues praising the US and its potentialities by inviting Lalita's father to 'join' him to start a business (using a deictic verb implying movement), highlighting the advantages of living 'there' – where the local adverb stresses a distance that is not only spatial, but also social and cultural: the Indian community there is described as professional in opposition to the 'these uneducated' Indians, the demonstrative adjective unmistakably identifying the space which they are associated with. The only difference existing between the original and the dubbed version is probably in Lalita's words 'your US', expressing her annoyance at the representation of the US that Kholi is giving, which is lost in the Italian translation 'gli Stati Uniti'.

The next sample is part of the same scene.

Source text	**Target text**
Kholi: There's only one problem with America. <u>Our</u> girls that are born <u>there</u>, they've totally lost their roots. Completely clueless.	Kholi: C'è solo un problema in America. Le indiane, nate <u>là</u>, hanno perso le loro tradizioni. Sono delle degenerate.
Mrs Bakshi: <u>Our</u> girls are very rooted. They're really very, very, traditional, huh?!	Mrs Bakshi: Le nostre bambine invece no. Sono molto, molto, molto tradizionali!
K.: You know, <u>in the US</u>, they're all too outspoken and career-oriented and some have turned into …the lesbian! So that leaves me with no choice. I may be healthy, wealthy, and wise, but as they say, "No life…without a wife!"	K.: Sa, <u>negli Stati Uniti</u> sono troppo disinibite e pensano soltanto alla carriera e alcune sono diventate persino…lesbiche! Così non mi resta altra scelta. Io sono sano, ricco e saggio, ma come dice il detto "Niente sorte…senza consorte!"

Table 4. Geographical deixis in *BP*.

In Table 4, Kholi further emphasizes his belonging to a different community of Indians by using the possessive adjective and the usual spatial adverb, in 'our girls…born there', to mention a disadvantage of living far from India: although he has just applauded the Indian community in the US, he then criticises women because they have lost their roots, which does not seem to be regrettable when it happens to Indian men in the US. His words are repeated by Lalita's mother in reference to her own daughters who are, instead, 'very rooted' in what is now depicted as a positive space as far as traditions are concerned, India. The Italian translation does not feature this repetition, but explicitly identifies the referent by choosing to flatten it with 'le indiane, nate là', which underlines the implicit difference from 'Indian girls born here'.

What follows, Table 5, is an extract from another conversation between Kholi and Lalita, with most of the other characters also present and gathered in the sitting room of the Bakshi family.

Source text	**Target text**
Kholi: Ah, these Indians, they don't know how to treat tourists. There's no sophistication.	Kholi: Ah, questi indiani, non sanno accogliere i turisti, non sono attenti.
Lalita: What do you mean 'these Indians'? Are you no longer an Indian?	Lalita: Perchè dici 'questi indiani'? Tu non sei più un indiano?
K.: I'm a green card holder now. India is decades behind!	K.: Adesso sono residente in America. L'India è rimasta decenni indietro!

Table 5. Temporal and spatial deixis in *BP*.

By employing the deictic demonstrative adjective 'these', Kholi tries to distance himself from the Indians he is talking about (who cannot properly receive tourists), and that causes Lalita's objection: while ironically repeating his words, she reminds him that he is Indian too (despite the use of the temporal deictic element 'no longer').[9] Kholi then takes back her temporal reference and stresses that 'now' he is something else: he is 'a green card holder' (whose Italian translation is slightly more softened, 'residente in America'), which gives him a higher status compared to other Indians and to India as a whole that is 'decades behind', a further temporal deictic item.

9 Temporal deixis concerns the various times involved in and referred to in an utterance and includes adverbs (like *now*, *then*, *soon*, *yesterday*, *tomorrow*) and verb tenses.

4.2. *The Namesake*

Moving on to the second Desi film that this chapter discusses, *The Namesake*, this is Mira Nair's 2006 film based on the novel by Jhumpa Lahiri, and it depicts the life of Ashoke and Ashima Ganguli, two Indian first-generation immigrants who leave Calcutta and settle down in New York City. The film chronicles their cross-cultural experiences, those of their (American-born) children Gogol and Sonia, and their exploration of a multiple heritage straddling two worlds, India and the United States. The film was dubbed into Italian with the surprisingly non-equivalent title *Il destino nel nome*.

The following is the first scene of the film. Ashoke is on a train from Calcutta to Jamshedpur, reading the book *The Overcoat* by Nikolai Gogol, and he is repeatedly interrupted by another passenger, Gosh, just moments before a train crash.

Source text	**Target text**
Gosh: Seen much of <u>this world</u>? Ashoke: I went to Delhi once. Every year I visit my grandfather in Jamshedpur. G.: no, no, no, no, not <u>this world</u>. I mean <u>England</u>, <u>America</u>. I was in England for two years. It was like a dream. Sparkling, clean streets. Nobody spitting on the road. Have you ever thought of <u>going abroad</u>? A.: No G.: You should. You are young. You are free [in Bengalese with English subtitles]. I'll tell you one thing, you must <u>go out</u> and conquer the world. You will never regret it. A.: My grandfather always says that is what books are for, to <u>travel</u> without moving an inch.	Gosh: Sei stato un po' <u>in giro</u>? Ashoke: Sono stato una volta a Delhi. Ogni anno vado a trovare mio nonno a Jamshedpur. G.: No, no, no, non parlo <u>dell'India</u>, <u>dell'Inghilterra</u>, <u>l'America</u>. Ho vissuto due anni in Inghilterra, mi sembrava un sogno. Strade così pulite che brillano, nessuno sputa per strada. Hai mai pensato di <u>andare all'estero</u>? A.: No G.: Devi andarci, sei giovane, sei libero [in bengalese con sottotitoli in italiano]. Ti dico una cosa. Devi <u>viaggiare</u> e conquistare il mondo. Non te ne pentirai mai. A.: Mio nonno dice sempre che i libri sono fatti proprio per questo, per <u>viaggiare</u> stando seduti in poltrona.

Table 6. Geographical deixis in *TN*.

As shown by Table 6, spatial deixis can be said to be achieved in the English text through demonstrative adjectives, firstly, when Gosh talks about 'this world' having the whole world as a referent, which Ashoke does not recognize and mistakes for India (which then leads Gosh to specify 'no, not this world. I mean England, America'). The Italian version actually flattens such references by choosing the translation 'in giro', later clarifying what 'this world' stands for in its second occurrence, namely 'India' (which is never explicitly named in the corresponding English text and almost seems to be in opposition with the two other referents, 'Inghilterra' and 'America' in the Italian text). Besides the many toponymic references throughout the rest of the script, a number of deictic verbs are also frequently used to mark a movement away from Calcutta and India, something which seems to foresee Ashoke's future life far from that place.

In the following scene, Ashima, who now lives in America, is writing a letter to her relatives in India. While she wanders in astonishment through the streets of New York (where the sight of people washing their clothes – and even getting publicly undressed – at a laundrette causes her bewilderment), viewers are confronted with the sadness of her face which deeply contrasts with the (simulated) happiness of the words in her letter.

Source text	**Target text**
Ashima: I'm very happy, and everything <u>here</u> is wonderful. The plane journey was long but enjoyable, and I am feeling fine. […] The house is very nice, big, with a lot of windows. Do you know? <u>Over here</u> we get gas all day and all night.	Ashima: Sono molto felice e <u>qui</u> è tutto meraviglioso. Il viaggio in aereo è stato lungo, ma piacevole e io sto benissimo. […] La casa è molto carina, grande e con tante finestre, pensate, <u>qui da noi</u> abbiamo il gas tutto il giorno e tutta la notte.

Table 7. Geographical deixis in *TN*.

In Table 7, both the English and the Italian texts use a local adverb, 'here'/'qui', to refer to the new spatial location in which the character finds herself, after being almost thrown into another spatial and temporal dimension by a long plane journey from India. In addition, the use of the personal pronoun in 'over here we get…' reinforces the sense of her being part of another entity, America, where unlike India, she has 24-hour gas and drinkable tap water. As discourse unfolds and space changes, Ashima seems to shift from one identity to another, enacting a repositioning of herself by means of deixis.

In the next scene under discussion, Ashoke and Ashima's teenage children, Gogol and Sonia, are visiting their relatives in Calcutta. While Ashima feels at home among the many street noises, the yogis singing and the saris hanging out to dry, her son and daughter feel like foreigners.

Source text	**Target text**
Gogol: What's wrong with you? Sonia: I want to <u>go home</u>. G.: I have something that'll cheer you up. Here. The planner says that there are only 89 days left to go. G.: So what's wrong with you? Are you hot? Have some <u>ancient</u> air conditioning. S.: Go away.	Gogol: Ma che hai? Sonia: Voglio <u>tornare a casa</u>. G.: So io come tirarti su. Guarda, la mia agenda dice che restano soltanto 89 giorni alla nostra partenza. G.: Allora che hai? Senti caldo? Ecco, aria condizionata <u>vecchio stile</u>. S.: Vattene.

Table 8. Temporal and spatial deixis in *TN*.

In Table 8, the deictic item 'go home'/'tornare a casa' is used in reference to America rather than India – where Sonia's mother, Ashima, already feels at home. Linguistic deixis then highlights their belonging to different cultural communities and signals their acts of identity construction. This scene also offers an instance of time deixis, when Gogol makes fun of his sister using a fan as 'ancient air conditioning', symbolically contrasting two temporal dimensions having India as a referent for backwardness and America as a referent for technological progress.

Emotional implications clearly emerge from such deictic expressions that jointly hint at an opposition between the very concepts of space and place: whereas the former can be conceptualised in terms of location or extent, the latter involves the meanings that particular locations take, including the memories and feelings attached to them. In this case, within the Ganguli family, parents and children have antithetic perceptions of space and place, something which is linguistically marked by their use of deixis: whereas India is the parents' place, it is home to them, and America is just a cold and alien space, the children construct India as an unwelcoming and feudal space, and experience America as their place, which they call home. By investigating the various deictic forms used in the text (and assuming deixis as a mental space – as stressed by previous definitions), we are able to uncover the characters' different constructions of identity: changing spaces seem to correspond to changing identities.

5. Concluding remarks

The samples chosen from the films under investigation can be viewed as representative of the ways in which diasporic subjects position and distance themselves from the mother-country and from their host country through language. Analysis of the deictic expressions used in *BB* and *TN* has shown that linguistic elements are, indeed, imbued with specific cultural values pertaining to forms of identity construction, self-representation and self-identification. As discourse unfolds, the characters in both films perform their identities by shaping differing versions of India and the US. Although the chapter's spatial restrictions do not allow further examination – which is the reason why the number of films investigated was reduced to two – the filmic productions constituting the *DesiCorpus* can and do offer a variety of instances of speakers' positioning *in* language and *through* language.

While addressing the slippery notion of 'home', diasporic films are to be regarded as discursive texts and cultural products which shed light on novel modes of 'belonging' that cross nationality and ethnicity. By adopting the metaphor of the journey which unsurprisingly recurs throughout the *DesiCorpus*, diasporic productions seem to call for a repositioning of identities within the national space they inhabit. The meanings embedded in such films disclose a multitude of cultural resonances that shape identity as an ongoing process of negotiation. Although the system of representations is often articulated around traditional dichotomies (North vs. South, West vs. East, self vs. other), the language of diasporic cinema plays a crucial role in opening hybrid spaces of identification that seem to be indexically construed by both intercultural encounters and their resulting emotional relations.

The *DesiCorpus* of South Asian diasporic films may provide a platform for the analytical questioning of identity representation; thus, new critical perspectives on emerging linguistic and cultural formations, transcending borders that have become increasingly problematic, may contribute to the investigation of alternative forms of self-representation. By embracing both the 'hosting' and 'hosted' positions that are inherent to the very concept of 'hospitality' (Derrida 1997), diasporic films seem to become a shared *space* from which to make sense of phenomena such as migration and diaspora.

References

Anderson, James / Keenan, Edward 1985. Deixis. In Shopen, Timothy (ed.) *Grammatical Categories and the Lexicon.* Cambridge: Cambridge University Press, 259-308.

Balirano, Giuseppe 2009. Indian Diasporic Aesthetics as a Form of Translation. *Anglistica. Indiascapes: Images and Words from Globalised India* 12/2, 87-96.

Balirano, Giuseppe / Guzzo, Siria 2011. The Anglo-Italian Ethnolect as a means of Identity Formation in *Computer Mediated Cross-Communication.* In Bondi, Marina / Cacchiani, Silvia / Palumbo, Giuseppe (eds) *CLAVIER 09 - Corpus Linguistics and Language Variation.* Roma: Bulzoni, 137-159.

Benwell, Bethan / Stokoe, Elizabeth 2002. Constructing discussion tasks in university tutorials: shifting dynamics and identities. *Discourse Studies* 4/4, 429-453.

Benwell, Bethan / Stokoe, Elizabeth 2007. *Discourse and Identity.* Edinburgh: Edinburgh University Press.

Byrd Clark, Julie S. 2007. Discourse encounters through experiences at school: the notion of *Italianità* meets the construction of *la francité.* In Mantero, Miguel (ed.) *Identity and second lan-*

guage: Culture, inquiry and dialogic activity in educational contexts. Charlotte (NC): Information Age Publishing, 93-118.

Byrd Clark, Julie S. 2009. *Multilingualism, Citizenship, and Identity. Voices of Youth and Symbolic Investments in an Urban, Globalized World*. New York: Continuum Publishing Group.

Bucholtz, Mary / Hall, Kira 2005. Identity and interaction: a sociocultural linguistic approach. *Discourse Studies* 7, 585-614.

Cameron, Deborah 2000. *Good to talk? Living and working in a communication culture*. London: Sage Publications.

De Fina, Anna 2003. *Identity in Narrative. A Study of Immigrant Discourse*. Amsterdam and Philadelphia: John Benjamins.

De Fina, Anna / Schiffrin, Deborah / Bamberg, Michael (eds) 2006. *Discourse and Identity*. Cambridge: Cambridge University Press.

Derrida, Jacques / Dufourmantelle, Anne 1997. *De l'hospitalité*. Paris: Calmann-Lévy.

Fillmore, Charles 1997. *Lectures on Deixix*. Stanford: Center for the Study of Language and Information.

Freddi, Maria / Pavesi, Maria (eds) 2009. *Analysing Audiovisual Dialogue. Linguistic and Translational Insights*. Bologna: CLUEB.

Gambier, Yves / Gottlieb, Henrik (eds) 2001. *(Multi)Media Translation: Concepts, Practices and Research*. Amsterdam: Benjamins.

Gregory, Michael / Carroll, Susanne 1978. *Language and Situation: Language Varieties and their Social Context*. London: Routledge & Kegan Paul.

Hall, Stuart 2006. The future of identity. In Hier, Sean / Bolaria, B. Singh (eds) *Identity and belonging: Rethinking race and ethnicity in Canadian society*. Toronto: Canadian Scholar's Press, 249-269.

Hier, Sean / Bolaria, B. Singh (eds) 2006. *Identity and belonging: Rethinking race and ethnicity in Canadian society*. Toronto: Canadian Scholar's Press.

Lapadat, Judith C. / Lindsay, Anne C. 1999. Transcription in research and practice: from standardization of technique to interpretative positioning. *Qualitative Enquiry* 5/1, 64-86.

Le Page, Robert / Tabouret-Keller, Andrée 1985. *Acts of Identity: Creole-based approaches to language and ethnicity*. Cambridge: Cambridge University Press.

Lyons, John 1975. Deixis as the Source of Reference. In Keenan, Edward (ed.) *Formal Semantics of Natural Language*. Cambridge: Cambridge University Press, 61-83.

Malik, Sarita 2002. Money, Macpherson and Mindset: the Competing Cultural and Commercial Demands on Black and Asian British films in the 1990's. *Journal of Popular British Cinema: New British Cinema* 5, 27-45.

Manzo, Lynn C. 2003. Beyond house and haven: Toward a revisioning of emotional relationships with places. *Journal of Environmental Psychology* 23, 47-61.

Marimardou, Sophia 2000. *Pragmatic Meaning and Cognition*. Amsterdam: John Benjamins.

Naficy, Hamid 2001. *An Accented Cinema. Exile and Diasporic Filmmaking*. Princeton: Princeton University Press.

Rampton, Ben 2009. Interaction ritual and not just artful performance in crossing and stylization. *Language in Society* 38/02, 149-176.

Rossi, Fabio 2002. Il dialogo nel parlato filmico. In Bazzanella, Carla (ed) *Sul dialogo. Contesti e forme di interazione verbale*. Milano: Guerini, 161-75.

Scott, Mike 2008. *Wordsmith Tools* (version 5). Oxford: Oxford University Press.

Shopen, Timothy (ed.) 1985. *Grammatical Categories and the Lexicon*. Cambridge: Cambridge University Press.

Thomas, Jenny 1995. *Meaning in Interaction: an Introduction to Pragmatics*. London: Longman.

Part 2
Corporate and Organizational Discourse

JANET BOWKER

Multimodal, Virtual Professional Space: 'Unboundedness' and 'Grounding' in Corporate Webcasting Events

1. Introduction

It would hardly be an exaggeration to say that human communication in today's world is without boundaries, thanks to exponentially evolving technologies which are rapidly reaching out into realms once considered more germane to science-fiction, for example artificial intelligence and biological neuroscience. Infinite amounts of information are potentially available, in real time, to incalculable numbers of people, in spaces created by unlimited configurations of virtual platforms and data flows. The frontiers of time and space are being pushed forever upwards and outwards, and we have taken up full residence in Marshal McLuhan's 'Global Village' where 'the Medium is the Massage' (1967).

Yet, our boundlessness is not an empty space: it is grounded by a myriad of multi-purpose cross-networks achieving endless, timeless communicative endeavours.

> More people than ever collaborate or compete in real time with more other people, on more different kinds of work, from more different corners of the planet, on a more real footing, than in any previous time in the history of the world. (Friedman 2005: 8)

Space has become a hugely inhabited *place*. How virtual spaces are converted into purposeful places, and how this is achieved in and through language, is the subject of this chapter. The challenge it addresses is to investigate the relationship between channels and embodied discourse, to assess the match between their respective features, and to uncover the processes of 'grounding', the anchoring of 'unboundedness' (see Section 2 for full references) in the *here and now*. We shall see how the space is not just simply occupied, but is constantly re-made in a reflexive relationship with language. People orient themselves and their discourse to the medium's relevant features, while dynamically reconstructing the space itself, at the same time.

The material which has been selected to trace this belongs to one of the main drivers of global processes, partnering technological development, namely the field of CMBC, computer-mediated business communications, and I shall be investigating, in particular, corporate webcasting events.

2. The descriptive and explanatory framework: theoretical constructs, models, theories

2.1. *Mobility and stability: contextualization and relevance creation*

> [...] textual artefacts and multimodal semiotic aggregates become mobile mediational means [...] interactional contingencies change as a result of the mobility of participants and become fleeting and temporary [...] the here and now is transient and often unstable. (McIlvenny/Broth/Haddington 2009: 1880)

There has been a recent upsurge of interest in discourse studies and linguistic pragmatics in how "embodied actors communicate, interact and coordinate their activities in complex multimodal environments" (McIlvenny *et al* 2009: 1879; see also Goodwin 2000; Norris/Jones 2005). The above quotation reflects this 'mobility turn' in analysis and highlights the changing nature of 'interactional contingencies' imposed by the features of the medium: these have also been termed the 'affordances' and 'constraints' intrinsic to computer-mediated channels, as part of their 'compensatory properties' (Petroni 2011: 17). This means that while digitally-delivered communication has eliminated both time and spatial limits to the creation and dissemination of information, knowledge and ideas, it imposes all the more need for making this content accessible, salient and usable. This chapter deals with the discursive practices and the pragma-linguistic corollaries at work in these varied procedures of 'relevance making' and 'object definition' in virtual modes (Koschmann *et al* 2011: 521). These actions are transformational in this temporal-spatial nexus and, in themselves, constitute the processes of contextualization.

Crucial to the analysis is the application of the theoretical constructs of 'unboundedness' and 'grounding' (Pütz/Niemeier/Dirven 2001: xiv; Langacker 2001: 12 *et passim*).[1] The essential point is that both of these are not only part of the distinguishing characteristics of the channel itself – they are also key notions in cognitive linguistic theories and frameworks. This branch of linguistic investigation is therefore singularly equipped to illuminate these dynamics, and to explain how certain discourse features achieve their potential to meet the pragmatic goals described above. The use of cognitive linguistic explanatory models will reveal the closeness of fit between the virtual medium and embodied discourse, each mutually shaping and reenforcing the other.

The field of cognitive linguistics is increasingly seen to inform discourse, genre and modality analysis (Pütz/Niemeier/Dirven 2001 provide a detailed description of relevant applications of theoretical models). In the cognitive linguistic framework, language structure depends on and itself guides conceptualization: the latter is conditioned by our experience of ourselves, the external world and our relationship to that world. Linguistic units are therefore assumed to be abstracted from usage events, and consequently grammar is considered to be motivated by semantic considerations. In line with this way of thinking, grammatical markers such as tense morphemes are meaning-bearing elements, in the same way, in principle, as full lexical items (Pütz/Niemeier/Dirven 2001: xiii-xiv).

---

1 See, also, Langacker 2009 (148-180) for a comprehensive account of the very complex and many-faceted construct of 'grounding', and the related discussion of 'temporal coincidence' and 'epistemic immediacy' (Langacker 2009: 185-217).

2.2. Deictic syntactic constructions: profiling and viewing arrangements

Deixis is a central corollary of the cognitive linguistic assumptions described so far: with deictical referencing of person, place and time, expressed by pronominal usage and discourse markers of various kinds, for example, sense-making becomes dependent on understanding the user's perspective at a specific moment of utterance. The reverse is also true: this chapter, in fact, centres on how the proximal locational and temporal deixis of the 'here and now' is instrumental in the actual definition of speaker positioning. What is less appreciated, and possibly even less understood, generally speaking, is the deictic significance of the syntactic constructions of tense and aspect, and how these achieve their specific pragmatic significance.

To grasp this more fully, the fundamental cognitive linguistic principle of 'construal' is central, which refers to the ability to perceive and portray the same situation in alternate ways through language choice (grammatical class and derived morphemes, as well as lexis), which enables the participants to shift perspectives.

> Every lexical and grammatical element incorporates a particular way of construing conceptual content. It follows that a change in grammatical class involves a reconceptualization, an alternate construal resulting in a subtly different meaning in accordance with the abstract semantic values of the classes. (Langacker 2001: 8)

This 're-conceptualization' triggers changes in the 'viewing arrangements' created by grammatical choice and usage. Moreover, cognitive models consider spatial configuration to be functionally linked to the temporal plane – *now* is also *here* as in the common phrase *the here and now*; conversely, *not here* also means *not now*. So, in cognitive linguistic analysis, time reference is conceptualized and lexicalized in terms of proximal-distal (near and far) relations with respect to the experiencer. Tense and aspect usage, then, construct varying temporal-spatial 'profiles' (Langacker 2001: 9-12).

2.3. The 'internal view' of the present progressive tense: the 'locus' and the 'ground'

It is time to apply the previous theories and frameworks to specific instances of grammatical usage. In line with the notion that tense is a deictic phenomenon, Figure 1 below portrays the conceptual profiling, deictic centre and viewing arrangement created by the use of the present progressive tense.

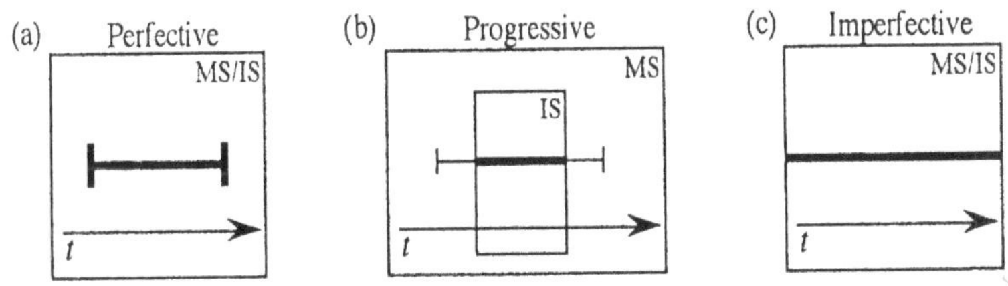

MS = maximum scope; IS = immediate scope; t = time

Figure 1. Conceptual profiling in the present progressive tense (Langacker 2001: 12).

Figure (a) describes the conceptual content of 'event', finite verbs, potentially having a beginning and an end point in time. Figure (c) illustrates the unbounded nature of stative verbs, verbs of mental state, of perception, and so on, conceptually unconstrained in temporal terms.

The present progressive tense, in figure (b), is analyzed in terms of a process of imperfectivization, that is to say it becomes unbounded, the end points being out of range and unimportant for this specific construal of events (the highlighted horizontal line demarcates the locus, the on-stage viewing). In fact, the present progressive tense is not distinct because of its focus on the present time but on account of the viewing arrangement it imposes. What counts in apprehension is the speaker's location of the process in the coding time, at the time of speaking. The deictic centre, the locus, the ground, in this case, is an internal view of an event (Langacker 2001: 21).

As a result, the use of the present tense is not restricted to describing events situated in present time. As we shall see in the analysis to come, the actual 'event time' is not foregrounded. This separation of the 'real' time from the 'discourse time', the time of speaking, this removal from time bounding, means that past, ongoing, or future events can be described using the present progressive tense, conjuring up this specific viewing arrangement, this 'internal view'. This emerges fully in the data in this study and is an important discourse feature of relevance creation.

2.4. Spatial imagery: the 'windowing of attention' and perceptual itineraries

One basic cognitive linguistic tenet, as we have seen so far in section 2, is that cognitive representation, accessing both perceptual and conceptual systems, consists of both language and visual space: Bloom *et al* (1996) have compiled a series of ground-breaking papers on varying aspects of this dense and challenging area. One of these areas of cognitive research, which identifies the links between perception and the sensory corollaries of language use, and which informs the analysis to come, is that of spatial and visual imagery.

Levelt (1989) has explored spatial and sensory (both visual and auditory) mappings in language, their contextual dependency, and the importance of linguistic deictic devices acting as pointers in the space created. Also, one of the founding fathers of cognitive semantics, Talmy (1996: 257), delineates a single, continuous cognitive domain which he terms 'ception', linguistic content having an analogue in visual format. Tenbrink (2011: 695) building on Talmy's seminal work on the 'windowing of attention' (2000: 257-309), adopts the premise that the dimensions of time and space together form a unified space-time continuum: she isolates motion and mobility as key features in temporal-spatial imagery, together with their fundamental attributes of extension, direction and transience.

All of these frameworks assist our understanding of the language use explored in this study. They help us to trace the processes of mooring and anchoring of discourse in an unbounded environment, in the depiction of

> […] an audio-visual scene which is more or less shared between the interlocutors, the places and orientations of the interlocutors in this scene, at the mo-

> ment of utterance and the place of the utterance in the temporal flow of events. (Levelt 1989: 45)

The dynamic scene that is created can then be filled with the embodied descriptive force which accompanies the use of sensory imagery. This is based on the idea that conceptualization is motivated by 'tight, recurring correlations in experience', termed 'experiential correlations' by Tyler/Evans (2001: 66). When using tense, then, we simultaneously express and apprehend the notions of salience, accentuation, actuality and intimacy, on a proximal-distal plane. This has significant bearing on the relevance creation procedures analysed in my own study, and allows us to view object definition not only in linguistic terms *per se*, but also to identify the role of subjectivization, evaluation and affective expression in foregrounding and relevance creation.

2.5. A cognitive view of the BE + GOING TO + VERB construction: grammaticalization and construal

A final area where cognitive linguistic models are of relevance to this study is in many ways an extension and summary of the rest of section 2, namely the use of the *be going to* construction in the expression of future meaning: there are links with the notions of the deictic significance of tense (2.3); the progressive aspect and its internal profiling at the moment of speaking (2.4); and motion in spatial-visual imagery (2.5).

This construction is an example of the process of 'grammaticalization' at work in the English language (Nicolle 2007: 47-65; Langacker 2009: 60-80) whereby lexical categories and constructions containing lexical material develop, in specific morphosyntactic contexts, into functional categories, including tense and aspect markers. Consequently the expression undergoes "semantic bleaching, in which components of meaning related to perceivable events and relations between entities is lost" (Nicolle 2007: 47).

In terms of conceptual profiling and construal, however, the proto-typical semantic sub-strata remain and can be described in a cognitive linguistic framework: "*be going to* features […] interpretation of the future as non-given, yet present, with a pending event's being signalled or announced at the time of speaking" (Brisard 2001: 251).[2]

Research indicates that, irrespective of the satellite of pragmatic semantico-grammatical meanings of *be going to*, the construction casts a profile, deictically, within the moment of utterance, anchoring the event to come (be it volitional, intentional, predicted, or whatever the functional connotation might be) in the present moment.

A final point of relevance to my study is that grammaticalization has been found to occur most often in constructions containing movement verbs, notably *come* and *go* (Nicolle 2007: 48) and that 'grammaticalization chaining' occurs due to increased functional dependence on associated lexical material (Nicole 2007: 47).[3] This helps to explain the high frequency of *be going to* in my data, and, together with its concordancing strings, clusters and contextual sequences, we can trace its importance in the process of relevance making, and the grounding of unboundedness in a discourse of the 'here and now'.

At this point, however, it is time to leave these linguistic abstractions, which nevertheless are fundamental constructs for an understanding of the analysis of this set of data, and to turn to the protagonists of the language events themselves, the spokespeople and stakeholders of corporate activity. In the next section we shall examine their own communicative purposes, the circumstances of their interactions and the means that are evolving to assist them in these endeavours.

---

2 Although the concept has become much refined and elaborated, thanks to cognitive linguistics, early linguists seemed to agree that *be going to* can be distinguished from most expressions of futurity in English by its relationship to current relevance. Leech establishes what was to be the forerunner of a proto-schematic definition for the entire semantic spectrum associated with *be going to*: "if there is one general meaning that can be attached to this construction, it is future fulfilment of the present" (1971: 54). Palmer, too, acknowledges this description, characterizing it as "current activity leading to a future event" (1979: 121).

3 This is testimony to the high degree of conventionalization of metaphor and the proto-typical spatial-visual imagery in grammatical usage, and helps to explain much of the troubled waters often connected with research into cognitive and linguistic metaphor.

3. Changing corporate priorities and communication resources

3.1. New exigencies and the knowledgeable organization

A number of discourse scholars have demonstrated that business communication priorities have been re-enforced and refocused in recent times, resulting in a proliferation of channels, resources and practices (Bamford/Salvi 2007[4], Bargiela-Chiappini 2009 provide a comprehensive compilation of current research orientations; Gillaerts *et al* 2012 have collated a wide range of recent research in business genres).[5] These exigencies are keeping pace (while also helping to set the pace, to a significant extent) with the far-reaching changes in the communication technologies of the 21st century, guiding and responding to deep socio-cultural change generally – in audiences, modalities, participation norms, and communicative intentions.

Firstly, successful business is increasingly dependent on a 'global reach' in order to achieve competitiveness. Effective corporate practice, then, relies on efficient knowledge management, both in external and internal dealings –the knowledgeable organization is indispensable for survival and success.

Secondly, organizations have become extremely fluid, permeable and are in rapid states of change. The need to manage infinite

4 This volume draws together the findings of the Rome Research Unit as part of a COFIN (later, PRIN) 2003-2005 national research project, comprising the University of Modena and Reggio Emilia, the University of Bergamo, the University of Florence, the University of Siena, and the 'Sapienza' University of Rome: the national project was entitled *Evaluation and linguistic patterning in written and oral business discourse.* The 'Sapienza' Group explored a wide range of business genres, modalities and types of organizational discourse, including transactional, promotional and expository.

5 The field of business discourse covers a wide-ranging and overlapping set of genres, settings and communicative events, but may be generally defined as "a process of talk and writing between individuals whose main activities and interests are in the domain of business and who come together for the purpose of doing business" (Bargiela-Chiappini 2009: 2).

amounts of information, instantly available in real time, makes learning a key issue in the creation, sharing and distribution of knowledge. This has led to radical changes in management-employee practices, prompted also by evolving work patterns and communications. An essential feature of this is multi-communication with multi-participation outside physical, geographical and temporal boundaries: this is the time of distributed work, multi-functional virtual teams, co-presencing and co-proximity, making use of sophisticated 'shareware' to make the connections 'real'.

This is giving rise to a new 'communicative agenda' in corporate communication, geared towards the timely, rapid and reliable sharing of information. These corporate efforts are increasingly aimed at building a persuasive image, identity and reputation, both for product and service branding, as well as cultivating a convincing internal 'employer brand'. This has always been the case, but now new exigencies are leading to different managerial discourses and styles in the achievement of communicative purposes. Increasingly business identities are multi-faceted, combining the professional with the personal, public 'face' with private 'face', and image-making aimed at displays of both competence and solidarity.

All of the above means that companies are concerned about being visibly more accountable to their stakeholders, both internally and externally: corporate style needs to demonstrate transparency and leadership, and consequently managerial discourse is increasingly geared towards the creation and maintenance of trust, credibility and confidence.

3.2. Evolving channels and modalities

These contingencies are giving rise to fast-emerging channels and media of communications, and producing evolving multimedial, multimodal formats and platforms for the achievement of the goals described above. A top executive from the company which provides the data used in this study had this to say about contemporary evolutions in communications and corporate objectives with reference to financial disclosures:

> The biggest shift in the corporate world regarding sharing the financials is the move away from annual and quarterly 'snapshots', looking back using written communications such as annual reports, memos, website communications, and into the audio-visual world. People like to *see* the CEO, *sense* his mood, *hear the stories*, to *read the message* behind the numbers. The main trend is a shift towards the blending of data, voice and video from a platform standpoint, and the opening up of two way interactive communications. This means accountability for what has not gone well, clear expectation-setting for the future, more regular updates on events affecting the financials, good and bad.
> (Author's italics for emphasis. Company executive, anonymous for confidentiality, recorded interview with researcher 2009)

Fast-changing web-casting events, incorporating video-conferencing techniques and state-of-the-art 'tele-presencing' encounters have emerged from the relatively rudimentary forms of audio-conferencing in use by international business not too long ago. Web-casting can be compared and contrasted with a number of genres and sub-genres in the business-reporting 'family', which in their turn are also developing and evolving: conventional annual reports, newsletters, mission statements and financial press releases are all downloadable and available to an unlimited audience. However, most of these are static, monologic and written. The interaction channel-modality features of web-casts, on the contrary, can be summed up as follows: they are increasingly

- multimedial, using telephone lines, computers and networking equipment in original ways and to special communicative effect.
- multimodal, using oral, written and visual configurations incisively and effectively.
- interactive and pro-active, prompting action and decision-making from the participants involved.
- aimed at multi-communicating: they are potentially unbounded in time and space, with multiple participants, potentially limitless audiences, in multiple locations.

Multimodal platforms, then, combine the features of 'unboundedness' and 'grounding'. There are, in fact, two temporal frames in their implementation: audio-conferences are relayed to live audiences, in a main office and networked out to others in synchronous, real-time reporting; they can then usually be accessed and re-accessed in an infinite time frame, through recordings and archiving. Reporting 'event time' is, then, disconnected in part from accessing time: although the event is rooted, moored in the 'here and now' it goes beyond the present: it is diaphanous, fleeting and due to its virtual nature, unstable. It can be re-contextualized and recreated in multiple time and place frames. The semiotic coordinates of the channel, then, consist of both contextualized and undefined boundaries.

4. The data, research focus and methodological approach

The business corpus used in this research is a multimodal one, a self-compiled, specialized set of company genres and sub-genres which I have been recording, transcribing and analysing over the last ten years or so. The data includes written, spoken and multimodal material and covers a wide range of participants, purposes and situations. The company is a North American-based world-player in organizational consulting. The corpus is divided into two main sub-corpora: in-house company communications, (oral presentations of company updating, financial reporting, 'webinar' in-service development and training sessions, e-newsletters) and external client solution consulting events (meetings, workshops, briefings, presentations and interviews).[6]

For the purposes of this study a selection was made from the in-house data: ten hours of webcasting in audio-conferencing mode, comprising around 150,000 words. The facilitators are all native speakers of either British or American English and occupy upper and middle-management positions. The audience is composed of employ-

6 To date the corpus has been used to investigate a number of topics and themes: the linguistic resources and pragmatic processes of description, evaluation and persuasion in corporate discourse (Bowker 2009); interactive networks of organizational metaphor and corporate identity (Bowker 2011); intercultural business encounters (Bowker 2012); genre and register blending and hybridization in multimodal, computer-mediated business communication (Bowker 2013).

ees throughout the Pacific Region of America (five states), from all ranks in the company. They are divided between a live public in the host office and those simultaneously connected (over the phone lines and computers) throughout the company's regional offices. The webinars are also accessible to potentially all interested parties for a limited time afterwards via the web. The research orientation has emerged throughout the course of this chapter, but more specifically the essential questions are:

- If 'unboundedness' and 'grounding' are particular features of virtual communicative space, how are these processes reflected and construed in discourse, and what is the relationship between the two?
- In terms of filling the space, and creating place, how are the pragma-linguistic resources of corporate-managerial discourse used in webcasting? In this professional virtual space, how does the company achieve its specific communicative purposes, related to audiences, referential relevance and interactional involvement?

In order to identify and track the creation of the 'here and now', this localized environment in a dispersed, mobile and transient one, we need to look at the processes of relevance-making, contextualization, subjectivization and personalization procedures as they emerge from the data itself. In order to do this, the following methodological tools are used in the analysis:

- Corpus linguistics software, particularly Wmatrix (Rayson 2003) for its USAS semantic tagging and POS (parts of speech) grammatical tagging functions.
- Interaction and conversational analysis, looking particularly at adjacency, sequentiality and discourse embeddedness.
- Ethnographic interviews and questionnaires with participants about the nature and value of their own corporate communications.

5. The analysis

5.1. The semantic profile: semantic fields and linguistic exponents

Using Wmatrix semantic tagging tools we derive a semantic profile for the oral presentation part of the data, as shown in Table 1 below.

Semantico/grammatical field: tag sets	Linguistic exponents
Pronouns	we, that, I, you, our, it, what, this, its, which
Quantities	some, number, numbers, half, %, per cent, net, margin
Sensory – sight	see, look at, looking at, seeing, seen, saw, looked at, watch
Boosters	really, very, more, so, a lot, much, very much, strongly
Evaluation – good	well, great, good, wonderful, ok, improved, terrific, fantastic, positive, improvement
Business – selling	clients, client, markets, sales, market, marketplace, billed, selling
Location – direction	this, here, out, there, where, transfer, top end, in, north
Cost and price	billings, billing, costs, cost, expenses, bill, worth, fees, charges
Present – simultaneous	now, to date, right now, today, update, so far, yet, at this point, going

Table 1. Key semantic fields and linguistic exponents.

In the top semantic fields, apart from the obvious categories pertaining to *business* (buying and selling, figures) *sensory-sight*, *location-direction*, *present-simultaneous*, *boosters* and *evaluation-good* are highlighted as the most prominent: these features were consequently selected for investigation using electronic retrieval tools, initially, followed by interactional analysis.

5.2. Concordancing strings for the proximal deixis 'here' and 'now'

Locational and temporal deixis was analysed through concordancing, collocational, collegational sequences and extended co-texts. Excerpts from the strings for *here* are shown below.

1 February results so what I'm showing	*here*	is an extract a couple a slides
2 last September so I will go public	*here*	and stick my neck out and say I
3 in fact we have three hundred clients	*here*	to date and over half of them
4 in October I just have the numbers	*here*	through September and you can see
5 big and 2006 you can't see this from	*here*	the actual projected OUP for 2006
6 looking that because you know he is	*here*	with us today and we're really
7 financial results through April and	*here*	we have a very nice picture to
8 and you have to put it in context now	*here*	if you look at the high point April
9 you know there are lots of themes in	*here*	about career development
10 so that's really my personal goal	*here*	to make sure that stays strong
11 seen a variety of things going on	*here*	with integrated solutions a real
12 of our billing clients for the north	*here*	again we see the usual subsets

From the examples it is clear that there is an almost indivisible divide between temporal, spatial and epistemic referencing: some lines are more obviously attracted to one area or another, but most reflect *the deep conceptual 'mix'* which enacts the discourse functions of *here*.

However, for the sake of simplicity, three main usage "attractions" or prominence could be distinguished from the data:

- Temporal: now, today (lines 1-3)
- Location: in this place, situation (lines 4-6)
- Epistemic: in this discussion, about this topic (lines 7-12)

Yet this would be a simplification, missing the point that the temporal-spatial-conceptual lines remain very blurred as the phraseology of *here* also shows: collocations of *here* include verbs of visual perception, *show*, *see*; fixed collocations and phrases, *here to date* = until now, *and here again* = and once more, *going on here* = happening, *now, here* = as a metadiscursive indicator and explicit pointer of attention. This is all indicative of the complexity of the referencing, grounding processes at work in the use of *here*.

The same conceptual blending emerges from an analysis of *now*, as the excerpts below indicate:

1 so if we focus on them we'll move on	*now*	to the financial results
2 which I have to say are looking terrific right	*now*	so congratulations everybody
3 than that I think I don't have the number	*now*	in front of me but you know
4 well there's a bit of a slow start but we're	*now*	seeing a lot of new activity
5 how that ties to our compensation results so	*now*	here is a shot of the latest
6 seeing results we do see the results here	*now*	this is not top line billings
7 score but its not as bad as the other areas	*now*	when you look at
8 so lets just get into some of the financials	*now*	so what I'm going to do is
9 if you look at the revenue but remember	*now*	this is after transfer in and
10 were billing less than 10 thousand a year	*now*	some of these we know are
11 both sides of this are running short of plan	*now*	you might say that's only
12 the same as 91% if we look at gross profit	*now*	gross profit is an important

The use of *now* is clearly multi-functional: it may refer more particularly to what the speaker is doing at that precise moment of speaking (lines 1, 3) indicated by phrases like *right now*, and *move on now*. It may also extend to ongoing events and states of affairs in the organization's business during that period, often used together with the present progressive tense (lines 2, 4).

But, as described in 2.4, *now* is very often unrelated to 'event time' but to 'discourse time', through the proximal perspective it creates: it has a fundamental role in the creation of discourse and making it relevant to the receivers at the time of speaking. Firstly it acts meta-discursively, managing interaction, directing the listener's attention and creating recipient involvement (lines 5, 8). Secondly it has an important textual function in argumentation, 'guiding and pointing' to the saliency of both upcoming information, cataphorically, and integrating this with what has been said, anaphorically (lines 6, 7, 9-12). In this it plays an important part in the achieving of understanding, promoting shared knowledge, and transforming the referential information to be presented into a personalized, assessed comment.[7]

7 The data has also revealed the close relationship of *now* with *so* (see Bowker 2013, for a comparison of the distribution, syntactic patterning and pragmatic functions of both *now* and *so* in the subgenres of power point slides, oral presentations and e-letters in corporate computer-mediated communications: distinctions between their adverbial functions and their meta-discursive, epistemic markings are indicated by the incidence of their mid-clausal as opposed to initial phrase position, and the presence of accompanying phonological signals, such as pausing and emphasis).

5.3. *Subjectivization and the proximal perspective: 'see', present tense, progressive aspect*

In the semantic profile obtained by Wmatrix, the *sensory-sight* category was highlighted. Indeed, the very high frequency of the verb *see*, and the salience of the sight and perception semantic tagging is far from coincidental – it plays a fundamental role in the creation of the descriptive discourse which is intensely embedded in the time of utterance (also frequently clustering literally with the words *here* and *now*, as these concordancing strings for *see* show).

1	profitability is really big for 2006 you can't	*see*	this from here the actual
2	advocacy motivation and satisfaction you can	*see*	here the results and as I
3	at the far right if you scroll down you'll	*see*	on the third line people are
4	as revenues grow over the next few years you	*see*	it's not a dramatic change
5	thank you for all your hard work and you'll	*see*	it coming up in the February
6	working so hard and not seeing results we do	*see*	the results here now today
7	and here we have a very nice picture to	*see*	and congratulations to those
8	business and as we get into discussing we'll	*see*	what we need to do
9	now this is not top line billings we have to	*see*	how the transfer in and out
10	I was mentioning to you the service advocacy	*see*	now if you look at that 94%

The close cognitive links between visual perception and conceptual representation (as revealed in everyday language in phrases such as *I see what you mean*, and *to see in your mind's eye*) also clearly emerge from the pragmatic configuration of the verb *see* in the data, and its subtly overlapping meanings:

- Sensory: literally visually perceive (lines 1-3)
- Cognitive: understand, know (lines 4, 5)
- Experiential: experience, have (lines 6, 7)
- Epistemic: consider, follow, investigate (lines 8, 9)
- Metadiscursive: now focus, think! (line 10)

Incidentally, also *nouns* of visual perception are common in the data and form semantic sequences related to sight: *a high level view, highlights, show-case, focus, line of sight, snapshot, picture, shot.*

The importance of the lemma *see* is also confirmed in the POS (parts of speech analysis) for verb frequencies and the derived tense morpheme -ING as shown in table 2 below.

VERB lemma	*normalized rel.%*	*-ING morpheme*	*normalized rel.%*
be	0.68	being	0.08
do	0.66	doing	0.24
see	0.55	seeing	0.13
think	0.54	thinking	0.06
look	0.41	looking	0.16
know	0.36	knowing	0.01
want	0.28	----	
get	0.21	getting	0.11
say	0.21	saying	0.10
need	0.18	----	
work	0.16	working	0.07
like	0.13	----	
talk	0.06	talking	0.03
come	0.06	coming	0.05
grow	0.06	growing	0.05
go	0.05	going	0.25
take	0.04	taking	0.04
run	0.01	running	0.09

Table 2. Most frequent verbs and morpheme -ING forms as normalized frequencies.

Verb usage in the data centres on the high frequency of statives, *be*; verbs of perception, *see*, *look*; verbs of cognition, *think*, *know*; volition, *want, need, like*; narrating, reporting verbs, *say*, *talk*; and highly frequent, very common action or event verbs, *get*, *work*, *come*, *grow*, *go*, *take, run.*[8] Given the importance of the description of figures and numerical data, contrary to expectations, apart from *grow*, the conventional lexical sets associated with describing trends and statistics are conspicuous by their absence: verbs such as *increase*, *decrease*, *rise*, *fall*, *climb*, *stagnate* are all but absent from the data, frequently with zero hits.

A different ranking, however, is obtained for the –ING tense morpheme: *going* 0.25, *doing* 0.24, *looking* 0.16, *seeing* 0.13, *getting*

8 This is clearly a feature of the spontaneous oral discourse and shows its lack of dependence on any power point slide verbal 'scripting' and its different relationship with non-verbal, visual displays in graphical form.

0.11, *saying* 0.10. Collocational analysis, together with co-text and extended contextual investigation, confirms the pragmatic value of the present progressive tense in creating the proximal perspective in relevance creation, as these extracts show:

(1) So what I'm showing here is an extract, a couple a slides from what X presented to the Board in December, going in explaining how we've done [...]. So we're seeing outplacement as a Transition business as probably being flat to 5% growth over the next few years and Consulting growth being 4 to 10%. Right now we're probably seeing like a 70-30 split of Transition to the other 3 buckets that's planned for in the Pacific Region [...].

(2) So we're at 96% through January and now I'll tell you how February looks – so for the region for the month – you know and the revenue is looking at 91% – and we are tracking pretty well on Transition and Consulting – about the same at 91% [...].

(3) So what does that mean for us? I think what it means for us is, we're not going to come out at 35% next year – I hope that we can set a bar that says we're not coming in at less than 40% and I think you should hold me accountable and the leadership accountable [...].

The first thing to notice is that the present progressive tense and related participles are used for a variety of temporal referencing, past, present and future, and not only for ongoing actions and events: *here*, *in December*, *over the next few years*, *right now*, *through January*, *February*, *for the month*, *next year*. This creates 'the internal view' described earlier (section 2.4), divorcing 'discourse', 'virtual platform' time from 'real', event referencing and allows the speakers to create a unique viewing arrangement, unbounding events in a sort of timeless present, and allowing them to hold them up for special attention at the moment of speaking.

This engagement strategy, anchoring referents in a virtual mobility, derives from the multimodality nature of the channel itself. The webcasts all make extensive use of power point type software or increasingly sophisticated 'groupware' or 'shareware' tools, providing a kind of lexical template in the form of strings of thought, a form of cognitive shorthand for the strictly-timed spoken exposition, and consequently restricting extempore speech. However, in data graphics, charts, graphs, diagrams and tables etc., the written language is absent and the presenter's job is to share the experience, visually, with the audience. As good data graphics are fully informative and self-explanatory, presenting complex quantitative ideas with clarity and simplicity, there is no need to do any duplication in speech of what is already comprehensively displayed on the computer screen — the presenter's role is to communicate a personalized, or, rather, a professional-corporate view of the data, transforming it into a prioritized evaluation, based on the selection and appraisal of information.

The above excerpts, in fact, illustrate how the presenter creates a subjectivized assessment of the value of the data for the audience, leading and pointing, transforming a static, object-centred graphical description into a dynamic, subject-centred one. This shift from the described to the describer explains how the participants use multimodal resources for their own communicative purposes: grounding processes become an entry point for corporate expression. In the above examples, presenters provide a concise summary of information, significant comparisons, incisive conclusions, and realistic predictions. They are at pains to create an image of corporate competence, demonstrated by company control of the situation, and communicated with transparency and realism. In this attempt to explain *what does this mean for us* and concede the need to *hold me and the leadership accountable* the presenter meets the audience's expectations *to see the CEO, hear the stories, read the message behind the numbers*, cited by the company executive in section 3.2, as being a significant element in current corporate identity construal priorities.

5.4. Visual-spatial imagery: evaluation and persuasion

The visual-spatial pragmatic effects deriving from deictic syntactic features also extends into the use of *be going to* constructions and associated imagery. *Going* is the most frequent -ING verbal morpheme in the data, as shown in Table 2. Moreover its use is almost equally divided between *be going on* (and several other, statistically much less significant prepositions: *before*, *through*, *from*, *after*, *in*, *off*, *back*, *around*, *up*) and *be going to*. A closer inspection of co-texts shows just how connected, pragmatically, the usage of *be going to* is, construing a multi-hued conceptual mix, similar to the complex proximal deixis of *here*, *now*, *see*, as the preceding analysis has shown.

(4) So today we are going to cover a number of things. Really we are going to go from a 30,000 foot level to a 7,000 foot level down to what's going on in our regions and our markets and it's a chance for all of us to get together and look at the challenges ahead.

(5) The Practice Leaders are really looking at their teams, finding ways in which we are really going to put the right people in front of the clients [...] we really have to think about how we are going to develop our people, so that's going to be our focus, what's going on in there.

The connection in the examples between *going on* (concurrently experienced) and *going to* (present expression of projected, unspecified future action) is the present grounding of events in the moment of utterance: a *going on* action is intimately related to what the company is *going to do*. A cognitive linguistic interpretation places the emphasis on the difference between the temporal-spatial profiling of the two prepositions (*on* indicating 'attached to', 'a part of here', whereas *to* points 'towards', 'from now'). In fact, any finalized past, present and future boundaries are superseded by the foregrounded salience of actions in the fluid mobility of ongoing communication.

This virtual movement, then, is created by the visual and spatial images of verbal use, focusing on motion, distance, location and direction. To finish, we shall see how all of the features treated so far become embedded in the discourse, textually interacting in a holistic pragmatic configuration. In the semantic profile obtained through Wmatrix described in section 5.1, the categories of *sensory-sight*, *location-direction*, and *present-simultaneous* emerged as being prominent, and the investigation of this data has revealed indispensable inter-correlations between all three. As importantly, the profile also includes *boosting* and *evaluation* in the top ten semantic-pragmatic areas. This final stage in the analysis enables us to integrate these components to the full, while at the same time bearing out the indispensable cognitive linguistic framework which has illuminated it, and without which the analysis would remain a mere collection of isolated and random discourse items. The following extracts illustrate this integration of linguistic and pragmatic features.

(6) We'll try to make this as dialogue and conversational as possible. We know it's a little hard when not looking at each other – we're all in our own spaces, virtually, all over the place. Now we're going to turn it over to Chris[9] and we'll jump into the webinar.

The presenter starts by admitting the constraints of the channel, and attempts to redress this by insisting on informality, setting the tone with a rousing *we'll jump into the webinar*.

The next extracts illustrate the widespread use of this visual-spatial imagery, intricately integrated with the deictic linguistic resources considered so far in the chapter.

(7) I wanted to show you this because it shows a realistic sense of how our revenues are growing over the next few years. You'll see it's not a dramatic growth that's projected. But if we contrast that with the next slide which is looking at OUP, there's a clear sense that we want to return to a good sense of profitability and 2008 you can't see this from here – the actual projected OUP for 2008 is 8% profit worldwide, going up to 13% and then 15 over the next couple of years. But that's really so, a modest growth but a much improved OUP [...].

The language of evaluation and boosting highlighted in the Wmatrix profile can be described using the notions of 'salience', 'accentuation', 'actuality' and 'intimacy', that is to say the 'experiential correlates' which accompany linguistic usage, elaborated in section 2.5 in the cognitive linguistic account.

Actuality, salience and accentuation feature prominently: *a realistic sense*; *a clear sense*; *a good sense* (semantic repetition and chaining is a commonly used resource); *the actual projected OUP*; *that's really so*. Not only, the visual reference is made explicit: *you'll see*; *the slide is looking at OUP* (in an interesting reversal of transitivity); *you can't see this from here*. The present progressive tense also maintains the proximal view, in significant departures from canonical grammatical description: *are growing over the next few years*; *is looking at*.

What this means in terms of identity is that the positive corporate image, the competence face being painted, needs to be credible,

9 Pseudo-names are used for all persons and companies for reasons of confidentiality.

and convincing. Company stakeholders need to be taken fully on board and consensus, cooperation and trust created.

(8) So how've we done this year? On the four areas you can see here the results. And as I have mentioned, people are passionate – this 94% on Service Advocacy is probably one of the highest you'll ever see in any survey anywhere around the world. But the contrast is how people felt about it at that time. "I would recommend the company as a great place to work" – 46% that's very low – that's horrifyingly low. So I have to say, when I saw this number I was just horrified and I thought "well what's going on" – and then I reeled back to what was happening last September. In the Pacific Region we had Global Rewards all being told that they're out of business and all of their friends feeling about this [...].

In this excerpt, boosting and evaluation are achieved through the adoption of a close perspective in terms of intimacy: the presenter is confiding with the audience, building relationships and communicating empathy and solidarity as part of employer-branding, creating the image of 'an employer of choice' (recorded interview with executive, 2007). The hyperbole reflects strong personalization and subjectivity: *passionate*; *horrifyingly low*; *I was just horrified*, imbuing the statistical information with a dramatic, affective force. Again there is a visual quality to the story, in this case almost cinematic, as the presenter *reeled back*, re-enacting the scene in direct speech. This is all part and parcel of Levelt's "dynamic audio-visual scene" created "at the moment of utterance and the place of the utterance in the temporal flow of events" described in section 2.5.

6. Findings and concluding remarks

This analysis of a corpus of corporate web-casts has revealed significant connections between modes, users and discourse in contemporary professional communications, in CMBC, computer mediated business communications, in particular. If we go back to the two research questions posed initially, we can say that regarding the relationship between modality and discourse there is a very close matching between their characteristics of unboundedness and grounding: the transcending of spatial and temporal boundaries of the channel is reflected in linguistic choices, especially through verbal imperfectivization processes: similarly, the need to contextualize within this virtual space on the part of users has a discursive equivalent in the use of proximal deictic resources and the language of evaluation and persuasion. The two, modality and discourse, are working together reflexively, each conditioning the other, reflecting their basic states of open-endedness, and the need to moor communication in the present moment. As such, discourse both arises from and moulds the channel: it is both the product and the re-enforcer of the modality.

Dealing in the main with the processes of contextualization, the study shows how these preferred linguistic forms and patterns actually work to establish locally-situated relevance in these mobile virtual environments, how they create continuity in discontinuity, connectedness in unconnectedness in a temporal-spatial dialectic.

As regards the second research question, the creation of corporate identity within this virtual professional space and its features, we have seen how users maximize on both the medium's opportunities and constraints, taking what was once local, 'in-house' communication, restricted to conventional audio-conferencing, out towards unlimited audiences and locations; at the same time, these long-range, professional voices are increasingly taking on a personal, subjectivized hue in the telling of 'the corporate story'. The public and the private spheres are becoming increasingly blurred in these fast-evolving channels in the meeting of new business challenges and exigencies.

The study also confirms what we know about the rapid evolution of genres within the virtual semiotic nexus, the phenomena of genre blending, hybridization and so on, where, however, traditional genre descriptive frameworks (fixed, pre-specified sequencies and stages, for example) have an increasingly uncomfortable fit. In the treatment of this corpus we have also seen the importance of powerful explanatory, descriptive linguistic models and theories in guiding and illuminating methods and analytical tools, in this case pertaining to the fast-developing field of cognitive linguistic pragmatics. As a final note, then, it may be useful, as part of a new research departure, to incorporate these into a reframing of classical register studies which goes beyond genre: this may result in productive conceptual frameworks which could account, instead, for what might well be termed 'web-registers' in virtual communication spaces.

References

Bamford, Julia / Salvi, Rita (eds) 2007. *Business Discourse: Language at Work*. Roma: Aracne editrice.

Bargiela-Chiappini, Francesca (ed.) 2009. *The Handbook of Business Discourse*. Edinburgh: Edinburgh University Press.

Bloom, Paul / Garret, Merrill, F. / Nadel, Lynn, F. / Peterson, Mary A. (eds) 1996. *Language and Space*. Cambridge (Mass.): M.I.T. Press.

Bowker, Janet 2009. Descriptive processes in business audio-conferencing: Telling the corporate story. In Radighieri, Sara / Tucker, Paul (eds) *Point of View: Description and Evaluation across Discourses*. Roma: Officina Edizioni, 175-193.

Bowker, Janet 2011. Crossing national business borders: 'doing' US management consulting in a global context. In Salvi, Rita / Tanaka, Hiromasa (eds) *Intercultural Interactions in Business and* Management. Bern: Peter Lang, 187-212.

Bowker, Janet 2012. From Communities of Practice to Communities of Learning: interdiscursivity in changing corporate priorities. In Gillaerts, Paul *et al* (eds), 115-138.

Bowker, Janet 2013. Variation across spoken and written genres in internal corporate communication: multimodality and blending in evolving channels. In Bamford, Julia / Diani, Giuliana / Cavalieri, Silvia (eds) *Variation and Change in Spoken and Written Discourse: Perspectives from Corpus Linguistics*. Amsterdam: John Benjamins, Chapter 3.

Brisard, Frank 2001. *Be going to*: an exercise in grounding. *Journal of Linguistics* 37, 251-285.

Friedman, Thomas L. 2005. *The World is Flat: A Brief History of the 21st Century*. New York: Farrar, Straus and Giroux.

Gillaerts, Paul / De Groot, Elizabeth / Dieltjens, Sylvain / Heynderickx, Priscilla / Jacobs, Geert (eds) 2012. *Researching Discourse in Business Genres: Cases and Corpora*. Bern: Peter Lang.

Goodwin, Charles 2000. Action and embodiment within situated human interaction. *Journal of Pragmatics* 32/10, 1489-1522.

Koschmann, Timothy / Le Baron, Curtis / Goodwin, Charles / Feltovich, Paul 2011. 'Can you see the cystic artery yet?' A simple matter of trust. *Journal of Pragmatics* 43, 521-541.

Langacker, Ronald W. 2001. Cognitive linguistics, language pedagogy, and the English present tense. In Pütz *et al* (eds), 3-37.

Langacker, Ronald W. 2009. *Investigations in Cognitive Grammar*. Berlin: Mouton de Gruyter.

Leech, Geoffrey N. 1971. *Meaning and the English Verb*. London: Longman.

Levelt, Willem J. M. 1989. *Speaking: from Intent to Articulation*. Cambridge (Mass.): M.I.T. Press.

McIlvenny, Paul / Broth, Mathias / Haddington, Pentti 2009. Communicating place, space and mobility. *Journal of Pragmatics* 41, 1879-1886.

McLuhan, Marshal 1967. *The Medium is the Massage: An Inventory of Effects*. London: Penguin Books.

Nicolle, Steve 2007. The grammaticalization of tense markers: a pragmatic re-analysis. *Cahiers Chronos* 17, 47-65.

Norris, Sigrid / Jones, Rodney H. (eds) 2005. *Discourse in Action: Introducing mediated discourse analysis*. London: Routledge.

Palmer, Frank R. 1979. *The English Verb*. London: Longman.

Petroni, Sandra 2011. *Language in the Multimodal Web Domain*. Roma: Aracne Editrice.

Pütz, Martin / Niemeier, Susanne / Dirven, René (eds) 2001. *Applied Cognitive Linguistics I, II: Theory and Language Acquisition*. Berlin: Mouton de Gruyter.

Rayson, Paul 2003. *Wmatrix / USAS, Semantic Annotation System*. Lancaster: University of Lancaster, UK.

Talmy, Leonard 1996. Fictive motion in language and '*ception*'. In Bloom *et al* (eds), 211-276.

Talmy, Leonard 2003 [2000]. *Toward a Cognitive Semantics* (2 vols). Cambridge (Mass.): MIT Press.

Tenbrink, Thora 2011. The language of space and time. *Journal of Pragmatics* 43, 691-6994.

Tyler, Andrea / Evans, Vyvyan 2001. The relationship between experience, conceptual structure and meaning: The non-temporal uses of tense and language teaching. In Pütz *et al* (eds), 63-108.

CHRISTINA SAMSON

Lexical/Phraseological Representations of Identity: Corporate Performance and Risk Reviews

1. Introduction

In an increasingly competitive global economy, corporate organisations are keen to acquire a competitive advantage (Schwaiger 2004). It is, therefore, not surprising that building a corporate identity and reputation has become an important element in the strategic communication of companies (Dolphin 2004). Among the several communicative forms used to address share/stakeholders, annual reports are among the most significant as, on the basis of their informative role, companies attempt to create a strong and trust-based relationship with their readers. By doing so, companies establish a positive identity and build or preserve a reputation for themselves which is a crucial achievement, especially after recent unethical financial reporting practices.

As corporate identity and reputation are built on the information provided by the companies themselves or by other monitors (Devine/Halpern 2001), it has become essential for entities to follow specific financial reporting regulations which affect the type of information provided. These regulations attempt to establish the general purpose of financial reports which, as indicated in the International Accounting Standard Board (IASB) (2010), should focus on the financial position of a reporting entity, its economic resources, and the claims against the reporting entity as well as the effects of transactions and other events that change a reporting entity's economic resources and claims. Both types of information have to be useful, relevant and understandable to generate accountability in share/stakeholders. However, these financial reporting regulations are not mandatory. As a consequence, reporting narratives are either regarded as a means to provide biased information to mislead investors and/or stakeholders (Merkl-Davies/Brennan/McLeay 2011), or as a means to provide incremental useful information to improve company decision-making and profitability. Whether reporting narratives deliberately mislead investors or are effectively useful they, nevertheless, contribute to build the inherent character of the company, that is, its identity which is closely linked to its reputation.

Biased information on corporate activities is related to management strategies of obfuscation. Obfuscation is management's presentation of less desirable aspects of the organisation in as a favourable light as possible in the attempt to manipulate organisational audiences' perceptions (Brennan/Merkl-Davies 2012; Merkl/Koller in press; Brennan/Daly/Harrington 2010, to mention a few). Obfuscation has been widely and frequently investigated through, for instance, readability formulas (Sydserff/Weetman 2002), that is, the number of syllables and mean sentence length which are perceived to be good indicators of the syntactic difficulty of a passage to be read by a target audience (Bayerlein 2010; Courtis 2004; Rutherford 2003).

Useful information provided in reporting narratives is, instead, viewed in the literature as a form of impression management. The latter refers to the way managers (and/or their activities) present themselves to others in order to be perceived favourably (e.g. Hooghiemstra 2000). Impression management has mainly been examined through thematic manipulation analysis which enables the identification of rhetorical manipulations and of what is considered important for the reader/user. Thematic analysis allows us to understand how linguistic choices influence disclosure meaning (Brennan *et al.*, 2010; Yeung, 2007; Clatworthy/Jones 2003).

Other studies focus on corporate narratives as a proxy for corporate reputation (Craig/Brennan 2012), organisational identity or management accounting (Abrahamson/Englund/Gerdin 2011), to mention a few. All these studies are characterised by being one-dimensional, in that they apply only a quantitative or a qualitative methodology. They also focus on pre-selected index items and de-contextualise textual analysis, that is, they ignore the relation between narrative and context.

By contrast, there is a paucity of research applying corpus linguistics analysis to corporate identity and reputation. Nevertheless, pertinent studies to date include Bowker (2012) describing issues of 'employer branding' in internal organization communication; Malavasi (2010) investigating the multifaceted nature of banks' annual reports; Salvi (2007) discussing national identities and global culture of companies on line; Samson (2007, 2010) analysing the discourse style of companies boosting their identity in business to consumer (B2C) webpages and the differences between academic and corporate identities online.

Thus, in order to extend corpus linguistics analysis of corporate identity and reputation, this study investigates a specialised corpus of performance and risk reviews (PERISK) extracted from FTSE MIB (Italian Stock market) corporate annual reports. The methodology adopted is mixed. It starts with a Corpus-Driven approach which is then integrated with Discourse Analysis in order to attempt answering the following research questions:

- Which lexical items and phraseology recur most frequently across the corpus (PERISK)?
- How do the lexical items and phraseology encode corporate identity and reputation across PERISK?

The remainder of this chapter is structured as follows. The following section 2 focuses on the key concepts for this study, namely identity and reputation. Next, the data are described in section 3. Section 4 presents the methodology used in this research. On the basis of the methodological procedures, the findings are illustrated and commented on in section 5. In section 6, the key conclusions are outlined.

2. Identity and reputation

The concept of identity has undergone broad paradigmatic shifts over the centuries. For Taylor (1989), identity is a product of its historical conditions that is reformulated in strategic ways by the period under which it arises. From being a rarefied formulation of philosophers, identity has gradually been viewed as a self-fashioning project. More recent understandings of social and collective identity and according to postmodern accounts, identity is seen as fluid, fragmentary, contingent and essentially constituted in discourse. Thus identity is increasingly treated as something that is actively and publicly accomplished through discourse which, as Firth (1957) argues, cannot exist in isolation, detached from a specific environment. For instance, Benwell/Stokoe (2006) define identity in terms of how people display who they are to each other. Bucholtz/Hall (2005) argue that identity is the product rather than the source of linguistic and other semiotic practices and therefore it is a social and cultural rather than a primarily internal psychological phenomenon.

Thus, identity may be linguistically indexed through labels, implicatures, stances, styles, or linguistic structures and systems; identity, as Bucholtz/Hall (2005) argue, is relationally constructed through several, often overlapping, aspects of the relationship between self and other. Furthermore, identity may be in part intentional, in part habitual and less than fully conscious, in part an outcome of interactional negotiation, in part a construct of others' perceptions and representations, and in part an outcome of larger ideological processes and structures (Bucholtz/Hall 2005). The concept of identity may be extended to corporate identity wherein, drawing on Abrahamsson *et al.* (2011) at least two perspectives can be identified:

- A social actor perspective that is theoretically rooted in institutional theory, suggesting that identity resides in institutional claims and denotes a stable guide for how corporate members should behave and understand their organisation (Ravasi/Schultz 2006).

- A social constructivist perspective which argues that identity instead resides in collectively shared beliefs about the organisation, which are periodically renegotiated among the members.

In this chapter, the latter perspective is assumed, since identity is viewed as socially constructed understandings among corporate members that guides them as to what their obligations are, what is legitimate and how they are supposed to act in relation to external parties (Corley/Gioia 2004). Moreover, the constructivist perspective presumes that corporate identities are continually (re)constructed among corporat e members, and thus, as Whetten (2006) and Abrahamson *et al.* (2011) claim, are better described as fluid than enduring. This continuous (re)construction of identities depends on the social context in which corporate identity is narrated and through which, as He/Baruch (2010) maintain, reputation can be created, maintained or repaired by external communication, as for instance in impression management tactics and stakeholder information disclosure (Craig/Brennan 2012).

A reputation is, indeed, based on information about a firm's relative positioning in an organizational field, on marketing and accounting signals concerning performance, on institutional signals concerning social norms, on signals concerning firm strategy, and on attributes based on a firm's past actions (Schwaiger 2004). Thus, reputation, according to Devine/Halpern (2001) acts as a gauge of a company's sense of identity.

3. Data

The performance and risk reviews analysed in this study are the English versions of 26 Italian non-financial companies listed on FTSE

MIB[1] 2009. The decision to analyse reporting narratives of Italian listed companies responds to Merkl-Davies/Brennan's (2007) call for more international studies on accounting narratives which to date mostly focus on Anglo-Saxon managerial practices. Furthermore, performance and risk reviews are two key sections of Management Commentary (IFRS, 2010) for companies listed on the FTSE MIB, the primary benchmark Index of the Italian equity market. The index includes the leading companies across the Industry Classification Benchmark (ICB) sector in Italy, that is, approximately 80% of the domestic market capitalization.

The 2009 English version of performance and risk reviews was downloaded from the Internet to form a corpus (PERISK) which, for comparability reasons, was divided into two sub-corpora: Performance (436,395 tokens/words) and Risk (53,197 tokens/words) as shown in Table 1.

Activity Sector	*Performance*	*Risk*
BASIC MATERIALS	TENARIS	TENARIS
CONSUMER GOODS	BULGARI	BULGARI
CONSUMER GOODS	CAMPARI	CAMPARI
CONSUMER GOODS	FIAT	FIAT
CONSUMER GOODS	LUXOTTICA	LUXOTTICA
CONSUMER GOODS	PARMALAT	PARMALAT
CONSUMER GOODS	PIRELLI	PIRELLI
CONSUMER GOODS	TOD'S	TOD'S
CONSUMER SERVICES	AUTOGRILL	AUTOGRILL
CONSUMER SERVICES	LOTTOMATICA	LOTTOMATICA
CONSUMER SERVICES	MEDIASET	MEDIASET
HEALTH CARE	DIASORIN	DIASORIN
INDUSTRIALS	ANSALDO	ANSALDO
INDUSTRIALS	ATLANTIA	ATLANTIA
INDUSTRIALS	BUZZI UNICEM	BUZZI UNICEM
INDUSTRIALS	FINMECCANICA	FINMECCANICA

1 FTSE MIB is the main Italian stock market index of the 40 most-highly capitalized companies on the Borsa Italiana exchange. It was established on October 31, 2003, as the S&P/MIB index with a base value of 10,644, replacing the MIB 30 index. It was renamed the FTSE MIB index after FTSE took over running the index from S&P. (QFINANCE – The Ultimate Resource, © 2009 (http://www.qfinance.com/dictionary/ftse-mib).

INDUSTRIALS	IMPREGILO	
INDUSTRIALS	PRYSMIAN	PRYSMIAN
OIL&GAS	ENI	ENI
OIL&GAS	SAIPEM	SAIPEM
TECHNOLOGY	STMELECTRONICS	
TELECOMMUNICATION	TELECOM ITALIA	TELECOM ITALIA
UTILITIES	A2A	A2A
UTILITIES	ENEL	ENEL
UTILITIES	SNAM RETEGAS	SNAM RETEGAS
UTILITIES	TERNA	TERNA

Table 1. Performance and Risk sub-corpora.

The two sub-corpora include the following different categories of information:

- performance (performance indicators, financial position, investor information, group and subsidiaries performance, analysis of the activities in different areas of the organisation);
- risk (financial, strategic, operational, compliance, risk management structure).

The FTSE MIB companies in the corpora are classified according to their activity sector – Basic Material (BM), Consumer Goods (CG), Consumer Services (CS), Health Care (HC), Industrial (ID), Oil & Gas (OG), Technology (TC), Telecommunication (TL) and Utilities (UT). Table 1 shows that CG (27%) and ID (23%) form the majority of the companies included in the corpus whereas Utilities represent only 15%. Two companies from the Industrial sector were excluded: Fiat Industrial because it did not provide an annual report in English at the time, and Enel Green Power that provided annual reports only in Italian. Among the 26 companies forming the sub-corpora, Impregilo (ID) and ST Electronics (TC) did not provide risk reviews. Moreover, banks and insurance companies (12 companies) were excluded from the corpora, as their disclosures are significantly different from the industrial companies'.

4. Analysis

The analytical approach followed was a mixed one. It began with a corpus driven analysis in which the researcher is committed to the integrity of the data as a whole, and the descriptions of language emerge from the corpus itself (Sinclair 2004). The corpus is, therefore, not seen as a repository of examples that back pre-existing theories, but the theoretical statements are consistent with and reflect directly the evidence provided by the corpus. Examples are, therefore, not adjusted to fit pre-existing categories of the analyst; recurrent patterns and frequency distributions are instead expected to form the basic evidence for linguistic categories and the absence of a pattern is considered potentially meaningful (Tognini-Bonelli 2001).

4.1. Keywords, keyness

Wordsmith Tools 5.0 (Scott 2010), a commercial computer package, was used to generate two separate Word lists for comparability purposes – Performance and Risk. The Performance Word List and Risk Word List are illustrated in Table 2 and Table 3:

N	*Word*	*Freq.*	*%*	*Texts*	*%*
1	#	69,278	15.88	26	100.00
2	THE	29,806	6.83	26	100.00
3	OF	17,516	4.01	26	100.00
4	IN	12,058	2.76	26	100.00
5	AND	11,468	2.63	26	100.00
6	TO	9,030	2.07	26	100.00
7	FOR	5,986	1.37	26	100.00
8	A	5,632	1.29	26	100.00
9	MILLION	4,124	0.95	24	92.31
10	ON	3,657	0.84	26	100.00

Table 2. Performance Word list.

N	*Word*	*Freq.*	*%*	*Texts*	*%*
1	THE	4,136	7.77	23	100.00
2	OF	2,333	4.39	23	100.00
3	AND	2,141	4.02	23	100.00
4	#	1,603	3.01	23	100.00
5	IN	1,593	2.99	23	100.00
6	TO	1,518	2.85	23	100.00
7	A	741	1.39	22	100.00
8	GROUP	641	1.20	23	100.00
9	ON	592	1.11	22	95.65
10	RISK	540	1.02	23	100.00

Table 3. Risk Word list.

The Performance and Risk wordlists were compared with a reference corpus – DATA[2] – of about 892,908 tokens to create Performance and Risk Key word Lists, as shown in Table 4 and Table 5.

The key words are calculated by comparing the frequency of each word in the wordlist of each sub-corpus with the frequency of the same word/s in the reference wordlist. To compute the keyness of an item, the programme computes: its frequency in the smaller wordlist; the number of tokens/words in the small wordlist; its frequency in the reference corpus; the number of tokens/words in the reference corpus and cross-tabulates these. A word will be included in the key word list if it is unusually frequent (or unusually infrequent) in comparison with what one would expect on the basis of the reference corpus (RC), as can be seen in the Performance Key word list in Table 4 and in the Risk Key word list in Table 5 below:

2 DATA is a reference corpus which includes files of entire texts collected by the author during her participation in national interuniversity research groups sponsored by the Ministry of Education, University and Research of Italy. DATA is formed by files of published written economics lectures, industrial products, surgery products, EU and non-EU museum descriptions, collections, exhibitions, narrative guidebooks.

N	*Key word*	*Freq.*	*%*	*RC. Freq.*	*RC. %*	*Keyness*
1	MILLION	4,124	0.95	4,206	0.60	417.78
2	NET	1,714	0.39	1,847	0.27	136.46
3	YEAR	1,682	0.39	1,828	0.26	128.68
4	DECEMBER	1,429	0.33	1,506	0.22	125.64
5	SALES	1,095	0.25	1,152	0.17	96.94

Table 4. Performance Key word list.

N	*Key word*	*Freq.*	*%*	*RC. Freq.*	*RC. %*	*Keyness*
1	RISK	540	1.02	669	0.10	1,298.28
2	RISKS	354	0.67	463	0.07	825.39
3	GROUP	641	1.20	2,133	0.31	711.17
4	MANAGEMENT	219	0.41	617	0.09	288.76
5	AND	2,141	4.02	19,186	2.75	259.43

Table 5. Risk Key word list.

By comparing the Performance key words in Table 4 with those in the Risk key word list in Table 5, a difference emerges between the two Tables. In Table 4, the relative most frequent lexical item is *million* (0.95) whereas in Table 5, *risk* (1.02) is the relative most frequent item in the Risk sub-corpus.

4.2. Concordances, semantic prosody

The Key word Lists allowed the application of *WST 5* Concord programme and the access to information about collocates of the relative most frequent key word in Performance and in Risk, that is *million* and *risk.* Table 6 shows the repeated collocations of the key word *million* in the Performance sub-corpus and Table 7 the recurring concordances of the key word *risk* in the Risk sub-corpus. The concordances indicate the semantic preference of the node word (*million* and *risk*, in this case) and the occurrence of two or more words within a short space of each other in a text. The concordances highlight the nodes' typicality or individuality/uniqueness (Scott 2010; Samson 2010) within the texts of the corpora.

N	*Concordance / Set Tag*	*Word #*	*Sent. #*
1	e transaction amounted to $280 **million.** Production from	161	4
2	24, 2009 amounting to â,¬3,070 **million** is equal to the p	389	10
3	consideration amounted to $940 **million** net to Eni. The t	505	12
4	enditures amounted to â,¬9,486 **million** to enhance assets	1,313	29
5	ploration activities (â,¬1,228 **million**) achieved a number	1,333	30
6	ent expenditures were â,¬7,478 **million** (up 16.3% from	1,406	32
7	al cash consideration ($940 **million**), 25% of which had	9,068	278
8	A gain amounting to â,¬100 **million** was recognized in	9,092	279
9	24, 2009 amounting to â,¬3,070 **million** is equal to the	9,216	283
10	consideration amounting to $280 **million** includes gas	9,451	291

Table 6. Concordance of key word million.

N	*Concordance / Set Tag*	*Word #*	*Sent. #*
1	the following: (i) the market **risk** deriving from exposure	2,092	26
2	modity prices; (ii) the credit **risk** deriving from the possib	242	45
3	interparty; (iii) the liquidity **risk** deriving from the risk t	749	57
4	quidity risk deriving from the **risk** that suitable sources	646	61
5	be available; (iv) the country **risk** in the upstream busi	3,396	78
6	business; (v) the operational **risk**; (vi) the possible evo	941	86
7	ies on financial risks. Market **risk** Market risk is the po	1,423	131
8	cial risks. Market risk Market **risk** is the possibility that	358	133
9	pany actively manages market **risk** in accordance with a	1,492	169
10	nducting finance, treasury and **risk** management operations	1,385	185

Table 7. Concordance of key word risk.

4.3. 'Million': achievement and reputation

In Table 6, the lexico-grammatical patterning of the node *million* is quite clear. The nouns, verbs and figures are related to capital invested, revenues, or payments which will mainly be directed outside the country. These actions indicate the company's presence on foreign markets as well as dynamism and accountability. The association between the node and its co-texts has a definite semantic prosody, that is to say, the spreading of connotational colouring beyond single word boundaries (Partington 1998) which, in this case, carries a positive load.

4.4. 'Risk': prevention and credibility

In Table 7 the node *risk* is closely linked to nouns such as market and credit and to the consequences of external events or exposure on markets that generate risky situations. These, however, are promptly dealt with by management.

The qualitative analysis proceeded in a second phase by analysing the most frequent clusters. These in the Performance and Risk sub-corpora form recurring phraseology referring to companies representing their identity while attempting to preserve their reputation.

5. Findings

To examine more in depth the semantic prosody of the concordances in Table 6 and in Table 7, a cluster search of the two sub-corpora was undertaken. The relative most frequent clusters of the keyword *million* in Table 4 – Performance keyword list – emerging from the Concordancer statistical counts per 1,000 tokens are shown in Table 8:

N	*Cluster*	*Freq*	*Length*
1	AN INCREASE OF	132	3
2	MILLION IN 2008	107	3

Table 8. Clusters of key word million.

The cluster AN INCREASE OF (*132)* was qualitatively analysed across the Performance sub-corpus from which the following examples were extracted:

(1) The company enjoyed considerable growth in 2009, closing the year with revenues of 304,129,000 euros. This amount, which reflects an increase of 24.3 percentage points, represents a significant gain compared with the previous year, when revenues[…] . (8_dias~1.txt)

(2) Electricity sold by the International Division in 2009 increased by 6,308 million kWh, mainly attributable to an increase of 2,258 million kWh in sales on the Russian market, mainly following the expansion of operations. (9_enel~1.txt)

In examples 1 and 2, the cluster *an increase of* refers to comparisons of company growth which are calendrically (Levinson 1983: 73) anchored to a specified time (*in 2009)* or a period of time (*the year; previous year)*. The anchoring of time, on the one hand, indicates accuracy in reporting information by co-occurring with nouns such as *gains, sales*, or *revenues*. On the other hand, the calendrical reference to time contributes to build up a grounded reputation since, drawing on Nguyen/Lablanc (2001), reputation is the result of past actions of a company, or the perception of an organization which is built up over a period of time and which focuses on what it does and how it behaves.

Corporate identity and reputation are further reinforced, in examples 1-3, by precise figures such as *24.3 percentage points, 2,258 million kWh, EURO167 million* which evoke with their surrounding lexical environment the 'goodness' of the company. This, in turn, has the significant function of drawing attention to the company' competitiveness which implicitly differentiates it from its competitors. The level of 'goodness' is also preserved when negative news is included in the communication. This, however, is presented as a consequence of general circumstances for which the company should not be considered directly responsible, as example 3 shows:

(3) […] decline in revenues from small plants (Legislative Decree 387/03 and Law 293/04). These effects were partially offset by an increase of EURO167 million in revenues from bilateral contracts and EURO 22 million in increased revenues from subsidies. (9_enel~1.txt)

The positive connotation of the cluster *an increase of* in (1-3) is further enhanced by evaluative adjectives such as, *considerable, significant*, *partially* and/or by the recurring lexical items such as, *gain, sales, expansion, revenues, contracts, subsidies* which through their collocation create a cumulative effect that recalls the outstanding performance of the companies. The semantic prosody also explains the causes of such important results achieved from the viewpoint of the management. This perspective is frequently underscored with the aim of indicating that the companies' priority is to maximize gains, profits while increasing the company's and management's reputation. To achieve the latter, the lexical items and the recurring phraseology in the performance reviews focus mainly on the way companies act in a series of financial operations which attain successful goals.

In Table 8, MILLION IN 2008 *(107)* is calendrically anchored to a specified time (*December 31, 2009; in 2008*) or period of time (*in the previous year*) which, as previously mentioned, has the purpose of establishing a historical value of the company that is closely linked to its identity. This is supported by transparency on previous year's *net sales, cash flows, net financial position* and specified amounts of capital *Euro 132.1 million, E1,375.8 million, Euro 35.7 million* which are introduced through comparisons, as examples 4-6 illustrate:

(4) December 31, 2009 net sales amounted to Euro 132.1 million, up 6.0% compared with Euro 124.6 million in 2008. The overall change compared to the figure for 2008 was dependent on the richer product mix [...] (18_pir~1.txt)

(5) Operating activities generated cash flows of E1,375.8 million, compared with E1,292.9 million in 2008. The increase in operating cash flow reflects the improved contribution from movements in working capital [...] (3_atla~1.txt)

(6) The net financial position is a positive Euro 35.7 million compared with a positive balance of 15.0 million in 2008. This positive change essentially reflects an overall improvement in the management of working capital. (17_pir~1.txt)

The accuracy of dates, the comparisons between present and past, the repeated references to sales, cash flows, and the numerical data have the function of backing the financial information provided in the performance review. Here the narratives focus on the financial position of a reporting entity, its economic resources, the effects of transactions and other events that change a reporting entity's economic resources and claims, as required by the IASB (2010) regulations. In addition, the detailed data contribute to underline the positive value of the performance which, drawing on Christensen (2002), projects an unambiguous identity of what the company is and stands for. As to this point, the narratives are characterized by nominalizations which convey an implicit evaluation of the *product mix, movements in working capital, working capital* that have been managed. The positive identity is further reinforced by the use of evaluative adjectives such as *richer, improved, positive* that underline the stance taken by the management in presenting corporate performance to its share/stakeholders, as highlighted by examples 4-6.

The way companies want their share/stakeholders to identify their commitment and efforts to achieve positive results is underscored even more in the Risk sub-corpus. The relative most frequent clusters of the key word risk (1.02) – Risk key word list – emerging from the Concordancer statistical counts per 1,000 tokens, are indicated in Table 9:

N	*Cluster*	*Freq.*	*Length*
1	THE RISK OF	39	3
2	INTEREST RATE RISK	32	3

Table 9. Clusters of key word risk.

The cluster THE RISK OF (39) and its co-text encode the companies' effort to reduce or monitor risky situations by introducing appropriate management operational systems. Such an involvement of the companies is linked to the use of task-oriented verbs such as *introduce, reduce, define, design, monitor, categorise*, as illustrated in examples 7-9, which also assume a value laden shade of meaning, drawing on Malavasi (2007). For example, the repetition of the cluster *the risk of* is closely linked to the attempt of introducing protective systems which will prevent or reduce various forms of risk and fraud while implicitly expressing an evaluation of the various corporate decisions or plans, as expressed in example 7:

(7) The Company has introduced custom management systems and new SAP transactions aimed at reducing the risk of fraud and has defined specific

procedures for a number of critical processes designed to prevent such illegal acts. (25_ter~1.txt)

By underlining the consolidated monitoring functions of their operational activities, companies not only foreground their experience, but also reassure their share/stakeholders as, for instance, in example 8:

(8) The quantification, dissemination and operational monitoring activities can be considered consolidated and effective in monitoring the risk of non-compliance with the applicable regulations, also as a result of specifically designated corporate functions. (16_med~1.txt)

Pointing at the company's experience is so important that in example 9 the risk of work-related accidents has already been categorized. This has the function of reassuring share/stakeholders:

(9) Certain Group activities are, by their nature, more exposed to the risk of work-related accidents such as operational services on the territory and the performance of technical services. We have categorized them. (1_a2a_~1.txt)

All three examples 7-9 clearly indicate the strategic use of information in specific difficult financial operations which, on the one hand, entails underscoring how corporate management knows how to behave in critical situations, on the other, appears to be characterized by phraseology denoting the value of the company that diversifies itself from the others. Furthermore, the scarce use of evaluative adjectives show that the companies are attempting to apply reporting narrative regulations which call for company accuracy and reliability which entails a highly reassuring identity.

As the previous cluster, also INTEREST RATE RISK (32) collocates with task-oriented verbs such as *manage, minimize, achieve, organize.* These foreground corporate reliability and reputation which is based on the commitment to structural and operational values. For instance, in example 10 the company which had previously been involved in a financial scandal reassures its share/stakeholders that it knows how to easily manage interest rate risks:

(10) The Parmalat Group manages the interest rate risk with simple financial hedging instruments. (17_par~1.txt)

In example 11, the company is keen to inform its investors and stakeholders that it is already protected against risks and it has already planned to achieve precise financial objectives. The semantic prosody reveals the company's commitment to minimize risks in the delicate financial situation, even though evaluative language is not explicitly used. The prosody reveals what Martin (2000) terms as an 'irrealis' involvement of feelings related to future events:

(11) Interest rates affect the market value of financial assets and liabilities of the company and the level of finance charges. Eni's interest rate risk management policy is to minimize risk with the aim of achieving defined financial structure objectives. (10_eni~1.txt)

Similarly in example 12 the company decides how it has to organize its interest rate risk exposure with the aim of persuading its share/stakeholders of its future performance:

(12) To achieve this objective, the whole Group's interest rate risk exposure, in terms of notional amounts, must be organized as a mix of fixed interest rates and floating interest rates. (15_lux~1.txt)

All three examples 10-12 show that the companies are organized with different strategies to limit any risk which can arise from their financial operations. The high frequency of nominalizations – *interest rate risk, simple financial hedging instruments, interest rate risk management policy, defined financial structure objectives, interest rate risk exposure* – as those also in examples 7-9 – *custom management systems, new SAP transactions, specifically designated corporate functions, work-related accidents* – conveys a technical connotation of the language which according to the established reporting narrative regulations should be unslanted and objective. However, the technical connotation also contributes to obfuscate information, in order to avoid revealing too much of the risks the company and share/stakeholders are running. In other words, the addresser's expression of thought, and simultaneously, attitude, is conveyed by the choice of not saying something explicitly (Bertuccelli Papi 2000).

Hence, the use of lexical and phraseological repetitions in the reviews are multifunctional. They denote apparently unslanted information which actually attempts to obfuscate critical financial situations while conveying the engagement and commitment of the companies to reduce risks. This in turn represents companies as reliable and solid organizational entities which thus 'mould' their addressees.

6. Conclusion

The quantitative and qualitative analyses of the corpus PERISK, specifically its two sub-corpora Performance and Risk reviews, add further insights to the literature focusing on corporate identity and reputation. The findings show how companies strategically use lexical items and phraseology to represent their identity while attempting to preserve their reputation. The corporate identity emerging from the data appears to derive from the perception of features that are deemed central, enduring and distinguishing. That is, the notion of identity is strictly linked to those features which are perceived as critical and, as such, according to Whetten (2006), define what the company must do to avoid acting out of character.

Specifically, in the Performance Reviews, the relative most frequent key word *million* and its recurring phraseology in the narratives point at constantly underlining solid corporate financial positions which are to be viewed as the effects of successful management decisions and transactions. These operations are expressed through the use of explicit evaluative adjectives and implicit evaluative nominal groups which refer to the increase of revenues, gains and growth. To reinforce the impact of the communication on share/stakeholders, detailed information is provided. This has multiple functions: it appraises corporate strategies; it makes information transparent while following specific reporting narrative regulations; it builds the identity of trustful entities which are measured through their reputation.

Moreover, identity and reputation are conveyed as historical values built over time, since the data underscore how essential it is for companies to (re)construct 'who they are' and communicate it to important others. By contrast, negative performance is simply minimised and/or immediately neutralized by emphasising the attainment of successful achievements.

The Risk reviews, unlike the Performance reviews, are typified by technical nominalizations which tend to obfuscate the information conveyed. The relative most frequent key word *risk* and its recurrent phraseology focus on the consequences of risky situations and on the measures companies have taken or plan to take to minimize risks. Although *risk* is the key word, few details of the risks faced by the companies are ever provided. The intention of management is to underplay the issues while implicitly fostering addressee confidence by moulding their opinion in favour of the company (Malavasi 2007).

Thus, by captivating the performance and risk reviews readership in a subtle and tactful way through the recurrent lexical items and phraseology, which are prevalently characterized by concealed evaluative strategies (in accordance with IASB regulations), companies and their management generate positive judgements and emotions in their share/stakeholders. The latter, in turn, identify such features with company reliability which implies a positive reputation. This, though, needs to be maintained, or repaired, otherwise it might be lost through inadequate financial operations and/or inadequate external communication which is pivotal when considering identity and reputation as collectively shared beliefs about an organization.

References

Abrahamsson, Gun / Englund, Hans / Gerdin, Jonas 2011. Organizational identity and management accounting change. *Accounting, Auditing and Accountability Journal,* 24/3, 345-76.

Bayerlein, Leopold 2010. Positive versus negative news: Readibility and obfuscation in financial reports. *Working paper, School of Business, Economics and Public Policy, Faculty of the Professions,* University of New England. <http://www.une.edu.au/business-school/working-papers/policy-governance/pagwp10-5.pdf>

Benwell, Bethan / Stokoe, Elizabeth 2006. *Discourse and Identity.* Edinburgh: Edinburgh University Press.

Bertuccelli Papi, Marcella 2000. *Implicitness in Text and Discourse.* Pisa: Edizioni ETS.

Bowker, Janet 2012. From Communities of Practice to Communities of Learning: Interdiscursivity in changing corporate priorities. In Gillaerts, Paul / De Groot, Elizabeth / Dieltjens, Sylvain / Heynderickx, Priscilla / Jacobs, Geert (eds) *Researching Discourse in Business Genres: Cases and Corpora.* Bern: Peter Lang, 115-138.

Brennan, Niamh M. / Merkl-Davies, Doris M. 2012. Accounting narratives and impression management. *The Routledge Companion to Communication in Accounting*, <http://ssrn.com/abstract=1873188>

Brennan, Niamh M. / Daly, Caroline A. / Harrington, Claire S. 2010. Rhetoric, argument and impression management in hostile takeover defence documents. *British Accounting Review*, 42/4, 253-68.

Bucholtz, Mary / Hall, Kira 2005. Identity and interaction: a sociocultural linguistic approach. *Discourse Studies*, 7/4-5, 585-614.

Christensen, Lars T. 2002. Corporate communication: the challenge of transparency. *Corporate Communication: An International Journal,* 7/3, 162-68.

Clatworthy, Mark / Jones, Michael J. 2003. Financial reporting of good news and bad news: evidence from accounting narratives. *Accounting and Business Research*, 33/3, 171-85.

Corley, Kevin G. / Gioia, Dennis A. 2004. Identity, ambiguity and change in the wake of a corporate spin-off. *Administrative Science Quarterly*, 49/2, 173-208.

Courtis, John K. 2004. Corporate report obfuscation: artefact or phenomenon? *British Accounting Review*, 36/3, 291-312.

Craig, Russell J. / Brennan, Niamh M. 2012. An exploration of the relationship between language choice in CEO letters to shareholders and corporate reputation. *Accounting Forum*, <doi:10.1016/j.accfor.2012.02.004>

Devine, Irene / Halpern, Paul 2001. Implicit claims: The role of corporate reputation in value creation. *Corporate Reputation Review*, 4/1, 42-9.

Dolphin, Richard R. 2004. The strategic role of investor relations. *Corporate Communications: An International Journal*, 9/1, 25-42.

Firth, John R. 1957. *Papers in Linguistics 1934-51*. London: Oxford University Press.

He, Hongwei / Baruch, Yehuda 2010. Organizational identity and legitimacy under major environmental changes: Tales of two UK building societies. *British Journal of Management*, 21/1, 44-66.

Hooghiemstra, Reggy 2000. Corporate communication and impression management – New perspectives why companies engage in corporate social reporting. *Journal of Business Ethics*, 27, 55-68.

IFRS Practice Statement Commentary. 2010. *International Accounting Standard Board* (IASB). London: IASB.

International Accounting Standard Board (IASB). 2010. *Conceptual Framework for Financial Reporting*. London: IASB.

Levinson, Stephen C. 1983. *Pragmatics*. Cambridge: Cambridge University Press.

Malavasi, Donatella 2007. Lexical nanalysis of implicit promotional devices in bank annual reports. *Les Cahiers de l'ILCEA*, 9, 171-84.

Malavasi, Donatella 2010. The Multifaceted Nature of Banks' Annual Reports as Informative, Promotional and Corporate Communication Practices. In Evangelisti Allori, Paola / Garzone, Giuliana (eds) *Discourse, Identities and Genres in Corporate Communication. Sponsorship, Advertising and Organizational Communication.* Bern: Peter Lang, 211-33.

Martin, James R. 2000. Beyond exchange: APPRAISAL systems in English. In Hunston, Susan / Thompson, Geoff (eds) *Evaluation in Text: Authorial Stance and the Construction of Discourse.* Oxford: Oxford University Press, 142-75.

Merkl-Davies, Doris M. / Brennan, Niamh M. / McLeay, Stuart J. 2011. Impression management and retrospective sense-making in corporate narratives: A social psychology perspective. *Accounting, Auditing and Accountability Journal*, 24/3, 315-44.

Merkl-Davies, Doris M. / Koller, Veronica 2012. 'Metaphoring' people out of this world: A Critical Discourse Analysis of a chairman's statement of a UK defence firm. *Accounting Forum*, <doi:10.1016/j.accfor.2012.02.005>.

Merkl-Davies, Doris M. / Brennan, Niamh 2007. Discretionary disclosure strategies in corporate narratives: Incremental information or impression management? *Journal of Accounting Literature*, 26, 116-96.

Nguyen, Nha / Leblanc, Gaston 2001. Corporate image and corporate reputation in customers' retention decisions in services. *Journal of Retailing and Consumer Services,* 8/4, 227-36.

Partington, Alan 1998. *Patterns and Meanings.* Amsterdam: John Benjamins.

Ravasi, Davide / Schultz, Majken 2006. Responding to organizational identity threats: exploring the role of organizational culture. *Academy of Management Journal,* 49/3, 433-58.

Rutherford, Brian 2003. Obfuscation, textual complexity and the role of regulated narrative accounting disclosure in corporate governance. *Journal of Management and Governance,* 7, 187-210.

Salvi, Rita 2007. The English of companies on line: national identity and global culture. Theoretical issues. In Salvi, Rita / Bamford,

Julia (eds) *Business Discourse: Language at Work.* Roma: Aracne, 9-45.
Samson, Christina 2007. Academic and Professional Website Identities: Anything to Share? *Quaderni del Dipartimento di Linguistica.* Firenze: Università di Firenze Unipress, 167-80.
Samson, Christina 2010. Boosting the company's identity. Point of view and promotional discourse style in a corpus of business to consumer (B2C) webpages. *Textus*, 23/1, 139-60.
Schwaiger, Manfred 2004. Components and parameters of corporate reputation – An empirical study. *Schmalenbach Business Review,* 56/1, 46-71.
Scott, Michael 2010. *Wordsmith Tools, 5.0.* Oxford: Oxford University Press.
Sinclair, John M. 2004. *Trust the Text. Language, Corpus, and Discourse.* London: Routledge.
Sydserff, Robin / Weetman, Pauline 2002. Development in content analysis: a transitivity index and DICTION scores. *Accounting, Auditing & Accountability Journal*, 15/4, 523-45.
Taylor, Charles 1989. *Sources of the Self: The Making of Modern Identity.* Cambridge, MA: Harvard University Press.
Tognini-Bonelli, Elena 2001. *Corpus Linguistics at work.* Amsterdam: John Benjamins.
Whetten, David A. 2006. Albert and Whetten revisited: strengthening the concept of organisational identity. *Journal of Management Inquiry,* 15/3, 219-34.
Yeung, Lorrita 2007. In search of commonalities: Some linguistic and rhetorical features of business reports as a genre. *English for Specific Purposes*, 26, 156-79.

ERSILIA INCELLI

Managing Reputation through the Discursive Reconstruction of Corporate Identity: A Phraseological Approach

1. Introduction

The primary goal of this chapter is to explore the discursive construction of space, identity and reputation in corporate discourse, in order to identify phraseological patterns which describe the language features, and to interpret the patterns produced in functional terms. The present study describes a research approach that combines corpus-linguistic and discourse analytic perspectives to analyze the discourse patterns in a small corpus of texts from two different discourse communities, namely the diamond industry and its opponents, non-governmental organizations (NGOs).

The chapter addresses the dispute that emerged between the diamond industry and its opponents in the wake of growing social concern for the role of 'conflict diamonds' in Africa over the period 1997-2011. With the image of the diamond industry heavily damaged and its legitimacy undermined, the threat of boycotts forced the industry to adopt a socially responsible attitude and new discourse strategies, in order to reconstruct a positive image in the eyes of its stakeholders. This type of narrative is symptomatic of many companies which, as the 21st century took root, found themselves having to conform to new societal norms characterized by changing socio-political environments, such as the growing 'green' voice, globalization and the increasing transparency of the network society (Castells 2000). Companies began paying serious attention to their social and environmental impacts, and this is particularly reflected in corporate *annual reports to society,* which form part of the corpus data base for this investigation. Thus, the main aim of the study is to analyze the language and narrative of these reports and the discursive constructs used by a corporation to frame its responsibility to the environment and community.

The approach implies a systematic search for a correlation between linguistic form and function in discourse. To carry out this task, the study relies on both qualitative and quantitative analyses. The quantitative part is phraseologically-driven in its observation of how recurrent sequences (phraseological units) have pragmatic discourse functions within contextual spaces, reflecting how speakers/writers select different patterns for different social and/or institutional purposes, according to different ideational positions and identities (Biber 2006).

2. Historical-social context

A brief word can be said here about the historical-social situation in which the texts were produced. The role of the *De Beers* company is particularly important, if only because today diamonds have become a synonym of *De Beers*. Historically, *De Beers* created a cartel and the 'diamond mystique', building a reputation which spanned the whole of the last century. However, in 1998, Global Witness (an unknown NGO) published *A Rough Trade,* in which they publically targeted and accused *De Beers* of violating the UN sanction resolution 1173 by buying *conflict diamonds. Conflict diamonds* are rough diamonds traded by rebels to finance armed conflicts. In its accusations against *De Beers,* the report relied on the company's own Annual Reports and financial briefings, which openly boasted about their purchase of Angolan diamonds at advantageous prices. The NGO report portrayed the *De Beers* company as irresponsible, non–transparent and unethical, shaming *De Beers* for unacceptable corporate conduct. Initially the global diamond industry's response was minimal, but the Internet catapulted the issue onto the 'screen', and the 'innocent' image of the diamond, a symbol of *love, purity* and *eternity* became increasingly linked to war, and gruesome images, thereby threatening the economic position and reputation of the diamond industry. With increasing NGO and international pressure, the diamond industry realized that they could not continue to do 'business as usual'. There was a marked turnaround strategy and changes occurred in the company's rhetoric. The industry began inviting governments and NGOs to unite in efforts to solve the problem (the Kimberly Process will be mentioned in some examples below) . The increasing support of the company for cooperation with the campaign was seen by some NGOs as a shrewd corporate social responsibility strategy, and by others as another business strategy to maintain its disproportionate market share. Whatever the object of the diamond industry's motives for cooperation, it was probably the only way to regain credibility and restore consumer confidence to regain their reputation as a leader in 'ethical' trading. We shall now turn to how the discourse was constructed to do this.

3. Theoretical frameworks

3.1. A socio-constructionist approach

The discourse analysis component of the research is informed by social theory, and sees linguistic data as a social practice reflecting ideologies (Fairclough 2003). Insights from social constructionist theory (Berger/Luckmann 1967) provide a suitable approach to identifying the nature of organizational identities, perceiving organizations as linguistic social constructions, "continually constituted and reconstituted through dialogical processes", (Brown/Coupland 2004: 6, citing Boden 1994).

It follows that there is a substantial element of critical discourse analysis (CDA) involved, as it characteristically focuses "on the ways discourse structures enact, confirm, legitimate, reproduce or challenge relations of power and dominance in society" (van Dijk 2001: 353). Such an approach recognizes the strategic nature of discourses, that is, the ability for "discourse to be mobilized as a strategic resource", (Hardy/Palmer 1998: 1). As regards context, van Dijk (2005) emphasizes a 'socio-cognitive' approach to show how context is mediated through cognition, how individuals relate text and context through 'subjective mental models' (van Dijk 2005: 95). Fairclough (2003) recognizes here the importance of intertextuality in context, that is, how one text inter-relates with other texts. as well as the dialogic discourse resulting from the intertextuality of texts (Bakhtin 1986). This also involves knowing about the historical socio-political dimension in which a text is produced in order to carry out textual analysis (see section 2).

It is important to note here, however, that the CDA approach adopted in this chapter, understands 'critical' to incorporate both positive and negative language construction and deconstruction, as CDA already incorporates a positive element in its social argumentation (Flowerdew 2008). In this way, the approach hopes to exemplify a 'positive style' of discourse analysis (Martin 2004) focusing on the progressive shift from adversarial discourse to a discourse of mutual cooperation and collaboration.

3.2. Identity studies

From a socio-linguistic approach, the study explores the tripartite relationship between corporate identity, reputation and legitimacy, more specifically the reconstruction of reputation in terms of recreating identities. *Reconstruct* and *recreate* are used here to emphasize how reputation is constantly influenced by shifting perceptions of an organization's identity and behaviour. Positive reputation clearly generates an organization's goodwill in the public view, giving the company a competitive edge (Fombrum/Rindova 2000). Negative publicity, in today's 24/7 media coverage, can be a serious threat to a company's image and its social legitimacy. Therefore, the investigation is about how a company re-establishes congruency within its social space, between its values and acceptable societal practices, and how it constructs its corporate rhetoric to defend, maintain, restore reputation and trust. Reputation is understood here as "a perceptual representation of a company's past actions and future prospects that describes the firm's overall appeal to all its key constituents when compared to other leading rivals", (Fombrun 1996: 72). However, reputation cannot be understood without determining identity and, in turn, legitimacy. The concept of legitimacy, in this chapter, is taken as it is broadly used in management theory (Suchman 1995). Suchman (1995: 577) attributes three features to legitimacy; *cognitive* (in the sense of belief systems about the company and relevant events), *pragmatic* (related to the practical activities of the company and self-interest), and *moral* legitimacy (related to ethical judgment and viewpoints).

Regarding identity, in the past it was traditionally conceived as fixed and stable, confined to a company's logos and images representing the organization (Cornelissen/Elving 2003: 114). Today, identity construction can best be understood as a continuous process, evolving with ongoing globalization and technological progress, addressing eclectic audiences and shifting social perceptions. As a concept, it has become 'fluid and multilayered' (Garzone/Catenaccio 2009: 10), 'subject to multiple modes of construction and de-construction' (2009: 23). Organizations can now be characterized as having multiple identities, generated "in dialogic interactions with 'notional insiders' and 'outsiders'" (Brown/Coupland 2004: 3). This then, accounts for a company's engagement in 'on-going identity-centred debates' and in 'on-going argument' (2004: 3).

While the construction and representation of organizational identities has received considerable attention in the fields of both social sciences and linguistics (Spencer-Oatey 2007, *inter alia*), growing attention is now being given to identity-construction taking place in today's web-based spatial/virtual locations (Bowker 2013; Salvi/Turnbull 2007; Brown/Coupland 2004). Fewer empirical studies have focused on how companies construct their identities through dialogic processes of interaction with 'outsiders' or 'critics', or even opponents to the very identities companies have worked so hard at building. Furthermore, identity as intrinsically tied to legitimacy, although approached in management theory, has received relatively little attention in socio-linguistic analyses. Thus one of the objectives of the chapter is to investigate the link between practices with which companies acknowledge and address their responsibilities reflected in their organizational discourse, and how this discourse is used to gain legitimate consensus.

3.3. The research questions

In this light, in relation to discourse construction and language strategies, the following research questions have guided the analysis. How does a corporation, whose identity has been publically challenged, deal with the reconstruction of its image, identity and reputation, to convince stakeholders of its good intentions? How did the NGOs discursively expose and dismantle the diamond industry, so convincing society of corporate wrongdoings? As argument is a dialogic process and can only be understood via interactants, it was necessary to investigate the discursive construction of the NGOs' argumentative discourse, and the linguistic choices used to demolish corporate image. There is also the additional dimension of the dialogic relationship with the community in which the company operates, implying geographical and social space, and a third player in the overall discourse construction, namely, the local community.

To address these questions, a specific aim involved identifying highly evaluative recurrent language patterns, to capture the ways evaluative meanings are created and conveyed in the genre of company annual reports, in the multidimensional facets of corporate identity regarding self-promotion, corporate enhancement, corporate goodwill, corporate conduct, corporate culture and philosophy.

4. Corpus linguistics and phraseology

This study follows a body of research which adopts a framework combining discourse analysis (DA) and the quantitative techniques of corpus linguistics (Baker 2006). Partington (2006: 267) refers to what he defines "a nascent interdisciplinary field of corpus-assisted discourse studies (CADS)", in which the two approaches interact but remain informed by distinct theoretical concepts, implying different approaches to text analysis. The corpus linguistic (CL) approach taken in this chapter is informed by Sinclair's (1996) 'idiom principle', that is to say, the tendency for words to be co-selected entering "into meaningful relations with other words around them" (Sinclair 2004: 25).

Following the work of scholars such as Firth (1957), who claimed that the word in isolation was not the most useful unit of analysis when looking for meaning, Sinclair (2004) later expanded the scope of analysis to look at word combinations stretching the 'lexical item' to an 'extended unit of meaning' or 'meaning shift unit' (MSU), denoting that the 'co-selection of the constituents' causes a shift in the 'ambient meaning'. This unit of meaning takes account of other contextual factors consisting of five categories of co-selection: namely the 'core' (the node word); 'semantic prosody,' a concept akin to connotation (Partington 1998: 68), which connects crucially with lexical environment and provides clues to a word's function; 'collocation' and 'colligation' (grammatical choice) are related to the co-occurrences of the word and grammatical categories of the core word (Sinclair 2004: 32). Finally 'semantic preference', that is 'the restriction of regular co-occurrence to items which share a semantic feature' (Sinclair 2004: 142).

To briefly show how the co-selection of items works, we can take the example in the NGO texts of the verb *to fuel.* Retrieved extended units of *fuel* show that its most common collocates are *war, conflict, strife*, as in *Diamonds continue to fuel conflict, human rights abuses and terrorism.* In other words, there is a strong, negative semantic prosody attached to this word. Hunston (2002: 61, 141-142) refers to how semantic prosody can convey covert meanings, and hidden attitudes. In this case study there is no 'hidden' ideology, in that the texts are transparent enough in what they are about (and the wrongdoings of the diamond industry can now easily be searched on google). What is interesting is how the retrieval of recurrent units of meaning can uncover the different discourse strategies used to ascertain multiple identities (conveyed through linguistic and textual codification), which may not be easily discernible to intuition or the eye.

5. The corpus

A corpus of approximately 660,000 words (Figure 1) was compiled and divided into two sub-corpora; one consisting of twelve corporate annual *reports to society* over the period 2000-2011, downloaded from the *De Beers* company website[1]; the second consisting of a more mixed genre of 12 NGO reports, 23 press releases and 9 newsletters, spanning the period 1996-2011, accessed from NGO websites, namely *Global Witness, Survival International, Amnesty International.*[2] Annual reports were selected rather than corporate social responsibility (CSR) reports, because they are directed at a wider, eclectic audience and cover all aspects and categories of a company's identity. Although the NGO documents were a mixed genre, the communicative purpose and stance were the same, highly evaluative and argumentative.

Diamond Industry txt	**Opponents txt**
12 Annual Reports Total token/type = 419375/13980	12 NGO reports /23 press releases/9 newsletters Total token/type = 246584/13508

↓

Annual Reviews 2000 – 2005 Total token/type =143461/8550
Report to Society 2006 – 2011 Total token/type = 276564/10788

Figure 1. Total word frequency in the Diamond Industry corpus and the Opponents corpus. (*Txt* is used for filing purposes and refers to corpus)

As the analysis proceeded, it became apparent that knowing about the historical socio-political context was an important cognitive element to understanding what was going on, so as to monitor a valid interpretive analysis. I extended the corpus to other media and Internet sources on the issue of *conflict diamonds,* to objectively get a view from a wide angle and track the events over time. Thus, the *De Beers*

1 <www.debeersgroup.com>

2 The main document sources are from the *Global Witness* website: <http://www.globalwitness.org/conflict-diamonds>. The full list of documents can be obtained from the author on request.

corpus was further organized according to diachronic principles, providing a more complete picture of the evolution of the discourse (Figure1).

What was immediately apparent from the total frequency data was that although the Diamond corpus (totalling 419,375 tokens v. 13,980 types) was much bigger than the Opponents corpus (246,584 tokens v. 13,508 types), the type/token ratio was essentially the same. This advances the hypothesis that the language in the annual reports may be repetitive and perhaps more narrowly descriptive as a genre, than the NGO reports. Later observations, in fact, proved that the language of the NGO reports was more emotive, using a descriptively wider-ranging variety of words.

It is worth noting that the archives for *De Beers* annual reports (AR) on the website went only as far back as 2006, so I used the 'wayback machine', a web-tool which can retrieve former websites, to search for early annual reports[3]. Instead, the NGO websites allowed access to letters and press releases from 1996/1998, which made *intertextual* references to early *De Beers* AR and press releases not accessible on the *De Beers* website.

6. Methodology

Methodologically, this study combines the quantitative, empirical approach of corpus linguistics with qualitative discourse analytical methods (Baker 2006; Bondi 2007). The synergy between these two approaches allows a type of analysis which moves back and forth from the micro-level of the lexical unit, to the fuller co-text, mapping connections between meaning units and the text to make qualitative interpretations sustained by statistical evidence. Standard CL techniques were used, namely word frequency lists, concordance lines, cluster and collocational analysis, and concgram searches. These resources were available with the aid of three software tools, using de-

3 <http://archive.org/web/web.php> - Wayback machine tool.

fault parameter settings to generate normalization values: *ConcApp* (Greaves 2005), for the retrieval of frequency lists, concordance lines and key collocates; *Wordsmith tools* (Scott 1999) for word dispersion plots and cluster analysis; *ConcGram* (Greaves 2005) to extract recurrent concgrams, that is sets of between 2 and 5 co-occurring words, within a wide span (up to twelve words on either side of the node word), and which include all of a concgram's "configurations irrespective of any constituent variation (e.g. AB and A*B), and positional variation (e.g. AB and BA)", (Cheng *et al.* 2006: 414). Once identified the patterns were then investigated for their socio-pragmatic purpose and function in the specialized discourse. For example, the 3–word concgram *our/play/role* was observed for its pragmatic function in constructing institutional identity and spatial positioning on the market and in the community (see section 7.4.2).

As one of the tenets of corpus linguistics is that high frequency items in a corpus point to areas of potential interest, the initial stages of the analysis adhered to a corpus-driven approach (Tognini-Bonelli 2001: 84). This meant it was unnecessary to have prior research parameters, in that the retrieval of data is not pre-established, but rather the texts are left to speak for themselves as much as possible through data and statistical evidence. So the first analytical step was the extraction of single word tokens, followed by the retrieval of high frequency clusters (or ngrams) and phraseological units. Wordlists give us an idea of the topic or the 'aboutness' of the texts, or the style of a text (Scott 1999). For example, the much-used content words such as *diamonds, conflict, development, mining, community,* reflect the subject of the case study. Nevertheless, single words do not tell us enough about the semantic/syntactic relationships of the word items. More paradigmatic and syntagmatic information is derived from scrutinizing lexical items with their collocates within the wider co-text.

The study will focus on the following high frequency lexemes and lexical strings, obtained through diachronic and synchronic analyses:

- the recurrent lexical strings *licence to operate* (30 instances) and *triple bottom line* (9 instances), to illustrate how the narrative of the discourse changes according to corporate strategy. Although these phrases were not on the high frequency list, they often occurred juxtapositioned next to high frequency items such as *our, social, local, principles, operations* (section 7.2.1);

- the high frequency self-referentials, exclusive *our* (possessive adjective) and exclusive *we* (subject pronoun). These determiners are used as collective representations of the company, and produce patterns such as *our + nominalization + verb* (section 7.3);
- The collocations, colligations and semantic prosody around the high frequency items *community* and *role,* in particular the concgram pattern produced by *our/role/play* (section 7.4).

7. The analysis

7.1. Frequency lists

The analysis starts from a quantitative overview of the lexicon, identifying the most relevant topics across the corpora. In fact, the top frequency items correspond to different sections of the corporate annual reports, (closely related to the categories of a company's identity and to a company website), that is *mining, development, costs*; and to the main issue in the NGO reports, that is *conflict diamonds*. Table 1 shows the top 30/40 most frequent lexical items in each sub-corpus. The table is condensed ('zapped', to use *Wordsmith tools* terminology, Scott 1999), as high frequency prepositions and articles have been removed from the list.

Diamond txt				Opponents txt			
rank	token	tot. freq.	%freq./1000 words	rank	token	tot. freq.	%freq./1000 words
7	**our**	4343	0.9884 %	7	diamonds	3002	1.2174 %
11	De	3228	0.7347 %	8	that	2998	1.2158 %
12	Beers	3169	0.7213 %	9	diamond	2615	1.0605 %
17	diamond	2258	0.5139 %	18	**conflict**	1354	0.5491 %
18	that	2195	0.4996 %	19	**it**	1268	0.5142 %
19	**we**	2068	0.4707 %	20	be	1240	0.5029 %
23	US	1449	0.3298 %	21	have	1172	0.4753 %
24	million	1381	0.3143 %	25	**not**	958	0.3885 %
26	diamonds	1308	0.2977 %	26	trade	936	0.3796 %
27	mining	1300	0.2959 %	27	process	872	0.3536 %
29	development	1269	0.2888 %	29	industry	841	0.3411 %
30	**will**	1210	0.2717 %	37	**should**	445	0.1805 %

Table 1. Top frequency words in the *Diamonds* corpus and the *Opponents* corpus. (txt = corpus for filing purposes)

The most striking feature of the Diamond corpus is the saliency of self-mention, in the form of: the possessive adjective *our* ranked 7th (totalling 4,343 hits at a dispersion rate of 0.98% per 1000 words), followed by the corporate name *De Beers* (11th and 12th position) and the subject pronoun *we* in 19th position. Exclusive *our* and *we* function as self–referential determiners, referring to the *De Beers family of companies*, a collective term for the wholly owned and joint-venture business. Other high frequency key items are, *million, mining, development*, and *will,* all in the top 30. The main topics of the annual reports can be identified as related to 'finance' and 'diamond mining development'. Understandably *will* projects the company's short and long-term vision regarding strategy and planning. In the NGO corpus, the first content word is *diamonds* (7th position, totalling 3002 hits at 1.2% per 1000 words), followed by *conflict*, and other interest-raising items such as *it, not,* and *should.* These key items set the scene and contribute to what may be going on. For example, *not* as a very high frequency word (958 hits) is likely to have a powerful discursive effect in the negative stance of NGO rhetoric and world viewpoint; likewise the modal *should*, can be assumed to have the pragmatic function of admonishing the Diamond industry, saying what it *should be doing.*

As lists do not tell us the whole story, the next stage is to look at the concordance lines and collocates of selected items, which then require qualitative interpretation, admittedly in a monitored fashion. The extended co-text for the concordance patterns of *our* and *we* in the annual reports clearly show how they delineate the company's own sense of identity, as shown in the excerpt concordance lines in the following table.

1 we have the right leadership in place to lead **our** company forward. In July, **we**
2 What is the vision of the State Diamond Trader? **Our** vision is to create a business
3 on insulating our business from future shocks. **Our** strong balance sheet is the ke

Concordance 1. Sample concordances of *our* in Diamond txt.

In contrast, in the Opponents corpus *our* has a lower number of instances (109 at a dispersion rate of 0.05%), and acts differently as an identity modifier (Table 3).

1 issue for World Vision to prioritize into our advocacy and campaigning work,
2 its disastrous policy. We will not stop our campaign until the Bushmen are
3 civilians, I need to be clear about this. Our team was able to interview
4 co-nominated for the 2003 Nobel Peace Prize for our work on conflict diamonds.
5 on their purchase of rough diamonds, and to our knowledge they are not 46

Concordance 2. Sample concordances for *our* in Opponents txt.

Exclusive *our* refers to the type of work the NGOs carry out, rather than to institutional identity itself. It is used as a modifier for their expertise and knowledge in the *conflict diamond* issue, *e.g. our campaign/advocacy/website/team.* In effect the lexical string *to our knowledge*, in line 5 is recurrent (12 instances) in the NGO texts.

7.2. Diachronic analysis

Concordance analysis and a closer reading of the texts unveiled the narrative which was effectively taking place. The diachronic analysis provided more information on the evolutionary nature of the discourse, which in turn reflected different discourse strategies, according to the different stages in the dispute. Table 2 presents the keywords in the Diamond corpus for the two time-periods 2000-2005 and 2006-2011.

Diamonds Corpus 2000 – 2005			Diamonds Corpus 2006 – 2011		
rank	word	total freq.	rank	word	total freq.
17	our	713	7	our	3650
20	million	631	18	we	1575
26	we	502	30	family	919
42	Will	382	34	management	889
49	production	276	36	employees	874
50	Africa	274	45	local	782
54	mine	261	46	will	778
55	operations	260	52	social	654
56	development	258	54	community	620
57	directors	258	55	stakeholders	617
58	mines	255	57	Botswana	583
60	report	250	58	communities	582
62	financial	241	60	health	576
63	sales	239	65	society	529
65	management	235	66	Water	515
68	exploration	222	67	environmental	513

Table 2. Top frequency items for the Diamond corpus divided into two time-periods.

The key items in each time-period highlight different topics pertaining to each corpus. In the time-period 2000-2005, for the purpose and focus of this research, the top frequency words include *our, we, mining, production.* The focus is clearly on *operations, exploration, finance,* and *sales,* contrasting with the second period 2006-2011, which introduces 'new' high frequency items, such as *family, Botswana, employees, community, health, society, water*, as well as *stakeholders.* The focus in the latter sub-corpus appears to be on geographical space and place, involving *employees* and the *local* and *social community,* also evoking a nurturing/caring stance, reflected in the metaphor *family* and in the use of environment-related words. Closer text analysis confirmed the hypothesis. There is an obvious change of 'footing' in the discourse (Goffman 1981), from 2006 onwards. Wordsmith dispersion plots are particularly useful in highlighting the 'turnaround' discourse and implied 'reputation saving devices' which emerge. Table 3 presents the dispersion plot for the lexical item *community* in the full Diamonds corpus.

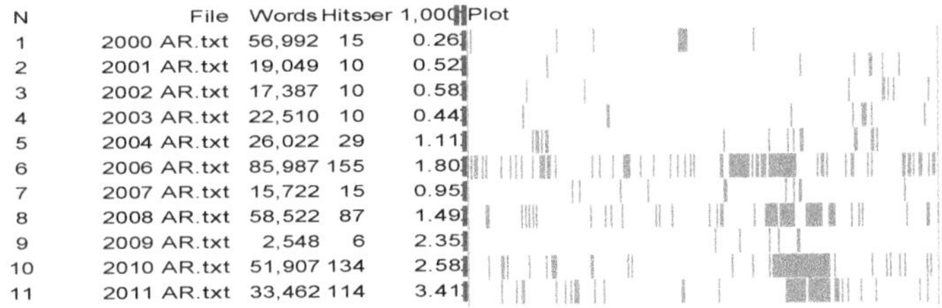

N	File	Words	Hits	per 1,000	Plot
1	2000 AR.txt	56,992	15	0.26	
2	2001 AR.txt	19,049	10	0.52	
3	2002 AR.txt	17,387	10	0.58	
4	2003 AR.txt	22,510	10	0.44	
5	2004 AR.txt	26,022	29	1.11	
6	2006 AR.txt	85,987	155	1.80	
7	2007 AR.txt	15,722	15	0.95	
8	2008 AR.txt	58,522	87	1.49	
9	2009 AR.txt	2,548	6	2.35	
10	2010 AR.txt	51,907	134	2.58	
11	2011 AR.txt	33,462	114	3.41	

Table 3. Dispersion plot for *community* in the Diamond txt.

It is clear from the plot that the item intensifies (in density and dispersion) in the annual reports from 2006 onwards. This coincides with growing awareness of the *conflict diamonds* issue and increasing corporate social responsibility (CSR) reporting.

The same retrieval procedure can be carried out for other words which present diachronic evidence in the texts, for example *sustainable, sustainability, land, local.* Graph 1 plots the frequency occurrence distribution for the items *community,* s*ustainability* and *sustainable* in *De Beers'* annual reports over the period 2000-2011.

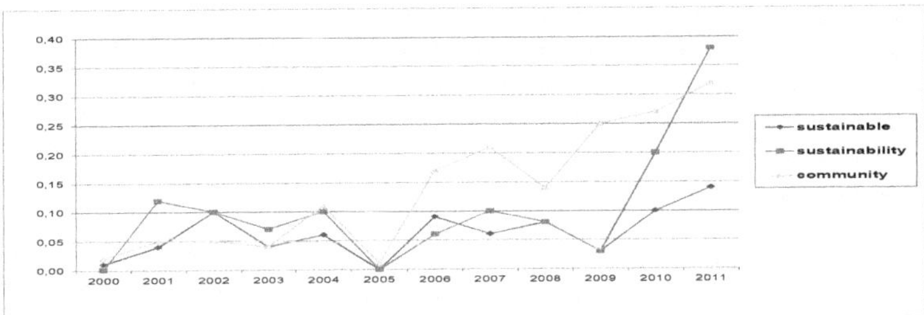

Graph 1. *Community/sustainable/sustainability* – diachronic shift in frequency use in *De Beers* annual reports over the period 2000-2011.

The graph shows the shift in the frequencies of the items over time, which is particularly prominent after 2006, reaching peak frequencies in the 2011 annual report.

7.2.1. Diachronic analysis of the lexical cluster 'licence to operate'

The diachronc analysis provided insight into the shifting discourse which the diamond company constructs to 'frame' its identity, and create alternative social spaces which justify its existence. This can be illustrated with two sample sets of extended units of meaning: *triple bottom line* and *licence to operate*. As expected, the conflicting discourse created by the *profits* and *principles* dictum, encompass an accounting framework that incorporates three dimensions of performance: social, environmental and financial, or *the triple bottom line*, such as in "*We subscribe to the concept of the 'triple bottom line' and seek to strike a balance between commercial success, environmental responsibility and social investment*" (AR 2000). However, interestingly, the string *triple bottom line* has 21 hits in the 2000-5 corpus, but no hits in the Diamonds 2006-2011 corpus. This may be indicative of how the company began using a different criterion in its spectrum of values for measuring its organizational success. If we observe concordance lines in Table 6 for the high frequency adjectives *social* and *local* in the Diamonds 2006-2011 corpus, we find that their strongest collocate is the lexical string *licence to operate,* which does not appear in the 2000-2005 corpus, possibly implying that explicit legitimizing strategies, as a form of discursive shaping and identity builder, began to emerge only after 2006.

```
1   ESTMENT we're more readily granted our local  licence to operate. That's true
2    a or cultural sensitivity. Securing a local  licence to operate relies on our
3  risks relate primarily to: ensuring our social licence to operate, conflict
4    social licence to operate.  Gaining a social licence involves ongoing
5 ommunity investment approach supports our social licence to operate by helping to
6    Me these standards helps maintain our social licence to operate and hence access
```

Concordance 3. Sample concordances for *local/social licence to operate* - Diamond corpus 2006-11.

The company's focus on social legitimacy is evident from the retrieved extended units of meaning, (Concordance 3) encompassing the verbs *ensure, establish, gain, grant, maintain, secure, support,* producing the performative pattern: *subject+verb+our local/social licence to operate.* The socio-pragmatic function of the above patterns are clearly to legitimize its spatial position within society and the community, to establish ethical accountability within its business approach. This is also indicated by the close colligate *our* as a self-mentioning determiner to which we will now turn for a more detailed discussion.

7.3. Focus on high frequency lexemes with 'our'

Changes in the discourse strategy can also be understood from *our* clusters and its collocates. A concgram search for the top 10 most frequent 2-word concgrams for *our* produced *our/to* (2049 hits ranking 4th position) and *our/we* (597 hits – 8th position), which were worth investigating for the purpose of this research. Concordance 4 presents a random sample of the prevailing pattern generated by the *our/to* concgram: *our + nominalization + to + nominalization/verb.*

```
1     areas. This commitment was made as part of our aim to minimise our environmental
2       impacts biodiversity directly because of our need to access, use and manage
3        De Beers. Managing Director's review 24 Our goal to grow our business to
4   out 99.8% of world rough diamond production. Our commitment to business ethics
5    ough the consequences of what we do so that our contribution to the world is
6     repositioning culminated in the setting of our strategic goal to achieve: "A
```

Concordance 4. Sample concordances for the *our/to* concgram in Diamonds txt.

Our patterns are at the interface of the company's self-appraisal, self-promotional discourse and planning strategy. The pragmatic discourse function of *our* patterns identify a variety of strategies for each aspect and category of corporate identity. I identified the most salient corporate strategies introduced by *our/to* and *our/we* concgram patterns, shown in the following excerpts from annual reports (2006-2011).

- Self-regulation and self-responsibilizing strategy

(1) The term 'living up to diamonds' – <u>our commitment to</u> operate in a manner that reflects the very qualities of the product <u>we</u> mine and sell – has become a shorthand for <u>De Beers' corporate soul</u>. It [...]. reflects <u>our values</u>, but also [...], <u>our commercial success</u>. (De Beers AR 2011)

(2) <u>Our approach to</u> sustainability is focused on managing and mitigating those risks that can affect <u>our commercial interests</u> by impacting on <u>our access to</u> supply or undermining consumer confidence. These risks relate primarily to: ensuring <u>our social licence to operate</u>, [...]. (AR 2010).

In examples (1) and (2), the rhetorical function of *our commitment to* and *our approach to* corresponds to a self-responsibilizing strategy used to enhance product image (*qualities, consumer confidence*), corporate conduct and reputation, ensuring *our commercial success.* As a commissive speech act *our commitment to* persuades the audience that the company can be trusted, because of the ethical way it presents its ideas, promoting company good will[4]. In addition, the company's persuasive rhetorical strategy of *ethos* and *pathos,* used to appeal to the audience's emotions through evaluative words such as *soul* and *values,* promote the *De Beers Family of Companies* as a caring corporation. Nevertheless, there is an element of ambiguity and tension in the discourse created by the juxtapositioning of *sustainability* with *our commercial interests*, *access to supply* and *ensuring our social licence to operate.* The company clearly has to maintain the support of a large stakeholder public, involving an eclectic mixture of customers, employees, stockholders, governments, as well as NGOs, so it is no surprise that the company appears to be doing a balancing act. Schuetz (1990) points out that sometimes companies structure communication to different audiences, but the result is "companies may advocate new policies and defend old ones in the same messages" (1990: 283). The intention is to legitimize their practices, but inherent tension remains between a discourse of intrinsic and instrumental values.

- Environmentally aware strategy

(3) Our actions to combat global change have two main workstreams of activity. We are committed to being a responsible energy user, recognised for our positive contribution towards combating climate change.

(4) In 2008, we revised our environmental strategy and framework[…]. The revised strategy reinvigorates our drive to align with the performance of the world's leading mining companies[…]to show leadership on biodiversity conservation.

4 Commissive speech acts, in which the speaker states a promise and an intention, are important in communication strategies as they can help reduce misunderstanding and conflict, as studied, for instance, in intercultural meetings (Bilbow 2002).

In examples (3) and (4) the rhetorical strategy is to evoke the image of an environmentally aware company, with 'modern' principles. The use of self-appraisal adjectives and evaluative nouns enhances their self-image, such as *positive, responsible, world's leading, leadership.* The company appeals to public ethos through the persuasive rhetoric of affirming the company's strategy of reflection, renovation and self-improvement, expressed through the determiners *our* and *we* colligated with the repeated use of *re*-prefixes, such as *revised* and *reinvigorate* in (4).

- Image repair and collaboration strategy

(5) This commercial imperative supports our determination to mine responsibly and underlines the importance of our work with industry, government and NGO partners to ensure that similar standards are maintained [...]. We have also actively worked with members to ensure the Kimberley Process [...] .

The strategy here is 'accepting responsibility' (*to mine responsibly*), and collaboration with *NGO partners* and *governments*. There was an obvious change of 'footing' in the discourse, which was no longer adversarial but collaborative, searching for a common ground. After such bad publicity from the *conflict diamonds* issue, any efforts to ensure regularity in the diamond industry would appear illegitimate without inviting the NGOs to negotiate standards.

- Leadership and marketing strategy

(6) We reaffirmed our position as the highest value diamond mining company in the world [...]. De Beers guarantees that 100% of the diamonds we sell are conflict free. All our diamonds are purchased in full compliance with national law, the Kimberley Process and our Best Practice Principles.

In (6), the company spatially positions itself on the diamond market as a world leader. What is more, the *our/we* concgram pattern reveals a savvy marketing strategy, i.e. *De Beers* guarantees that their diamonds are a *100% conflict-free.* C*onflict free diamonds* enhance the quality of their diamonds, and successfully rebuild consumer and stakeholder confidence, reinstating trust and reputation.

The overall aim of corporate communication in its annual reports is self-appraisal. The company is expressing its ability to establish congruence between the values implied by its actions and the accepted environmental and societal norms.

7.4. Corporate social and spatial positioning represented by heart/ role/community clusters

Table 4 presents the most frequent collocates of *our* in the Diamond corpus.

Diamonds txt - closest Right collocates for *our* = total 722					
rank	collocate	frequency	rank	collocate	frequency
1	operations	257	11	sustainability	56
2	business	171	12	Community	54
3	mining	101	14	Social	50
4	employees	100	21	environmental	37
6	Commitment	75	26	clients	31
7	Principles	74	27	safety	30
8	stakeholders	74	37	role	25
10	performance	58	41	heart	21

Table 4 . 'Zapped' right collocates for *our* in the Diamond corpus.

As we can see in the table, the strongest *our* collocates refer to the company's organizational activities and practices, for example *our operations/business.* Other recurrent collocates are *community, role* and *heart,* which are worth investigating more closely as spatial positioners in the texts.

7.4.1. Our/heart concgram

The following 2-word concgrams for *our/heart* (Concordance 5) metaphorically stress the company's business priorities, which are the *customer, diamonds, stakeholders, ethical accountability,* and *the community.*

```
1   sense of vision and responsibility are at the heart of our business model and
2  include ethical accountability which is at the heart of our approach to business.
3 partnerships with local communities sits at the heart of our approach to community
4   our stakeholders where we operate lies at the heart of our business philosophy.
5       customers have with diamonds  lies at the heart of the continued success of our
```

Concordance 5. Sample *our/heart* concgram – Diamonds txt.

The rhetoric produced by the pattern, *verb + at the heart of + our + nominalization,* appeals to the stakeholders' *ethos* and *pathos,* playing on the positive emotions associated with the *heart*, love, affection and care, along with the evaluative items, *shared, value, community, partnership, responsible,* etc. *Stakeholders,* the *community* and *heart* are inherently linked to the company's self identity and how the company positions itself within its social space.

7.4.2. Our/role/play patterns

The *our/role* concgram often appears in the company of the dynamic, action verb *play.* Concordance 6 presents a sample 3-word concgram search for *our/role/play* in the Diamond corpus.

```
1 confidence in our business and play a critical role in protecting the equity
2  disadvantaged groups, continues to play a key role in our approach to economic
3     South Africans will come to play an active role on the board of DBCM, while
4 to traditional values and a recognition of the role diamonds play as the gift
5  Leaders Forum, they recognise the significant role business can play in meetin
```

Concordance 6. *our/role/play* concgram – Diamonds txt.

We can see in the sample concordance lines above how the company places itself in a pivotal position on the global market, and in the community. Interestingly, an analysis of the 2-word concgram *role/play* presents insight into the semantic prosody created by the 'weighty meaningful' *role* played by the actors in each sub-corpus. Cheng *et al.* (2009) show how a configuration analysis of the constitutional and positional variation of a concgram string can illuminate meaning patterns, which contribute to the rhetorical and evaluative effect of the discourse. The *role/play* concgram was compared between the two corpora. In the Diamond corpus, 48 configurations were observed for the 2-word concgram *play/role* (which owing to space constraints cannot be displayed in detail here). Table 10 above presents a random sample including the most recurrent canonical form, i.e. *play*

** *role* with 25 instances, for instance *Diamond revenues play a key role in driving economic development in producer countries.* The benchmark pattern is *play + evaluative adjective + a role.* Other instances presented the following configurations: *play * role* - 5 times; *play *** role* - 3 times; *play *****role* - 5 times. On the other hand, the positional variant *role/play* had fewer instances: *role * play* - 5 times; *role ** play*- 3 times; *role *** play* - 1 time; *role **** play* - 1 time.

Following the concgram analysis procedure suggested by Cheng *et al.* (2009: 9), Sinclair's (1996) categories of co-selection can be identified for *play/role.* Its strongest collocations are the evaluative adjectives: *key, important, significant, major, leading, critical, central*; its strongest colligation is the modifier *our;* as to semantic preference we can identify nominalizations and compounds such as: *family of companies, business approach, organizational activities.* Overall, the semantic prosody of *play/role,* together with its modifiers and collocates, contributes to the 'weighty' role of the diamond industry and therefore it effectively contributes to the discourse strategies of self-enhancement, self-promotion, and the re-construction of reputation.

As a contrastive analysis, in the Opponents corpus the *role/play* concgram has a different semantic prosody and evaluative intention:

1 opean governments have a central **role** to **play** in strengthening the Kimberley Pro
2 the diamond industry, has a vital **role** to **play** in ensuring that self-regulation is
3 ates that "retailers have a crucial **role** to **play** in helping to ensure that the
4 ley Process. Consumers also have a **role** to **play** and should ask questions **to find**
5 not be purchased. This is the ideal **role** for the Security Council to **play** in such a
6 porfer con flict diamonds to **play** a **role** in fuelling armed conflict can be seriously
7 ing countries should **play** a central **role** in setting a regulatory framework within wh

Concordance 7. Sample *role/play* concgram in Opponents txt.

First of all, the *role/play* positional variant is more recurrent than the *play/role* one. The predominant patterns of the *role/play* concgram are: *subject/actor/agent + have + adjective + a role to play + in/to + verb*; or *the role that + subject/actor/agent + modal verb + play + verb.* The focus is on the 'negative' role the diamond industry is playing, and the 'potential' leadership *role* they *should* play, in terms of setting an example. The NGO texts use the *role/play* lexical strings to position their stance, and to persuade the diamond industry and the public, including *consumers, retailers, governments*, to take a stance in the conflict diamonds issue.

7.4.3. Corporate spatial positioning within the 'community'

Community/ties is a high frequency lexical item in the Diamond corpus, with a total of 987 instances (0.22% dispersion rate) v. 172 instances (0.08%) in the Opponents corpus. Table 5 presents the top 5 clusters for *community* in the two sub-corpora.

Diamond txt. Cluster	Freq.	Opponents txt. Cluster	Freq.
The local community	21	The international community	44
Community diamond partnership	21	The Lebanese community	9
Of the community	20	The Ngo community	5
Mwadui community diamond	20	Sense of community	5
Our community policy	19	A Bushman community	5

Table 5. Top 5 most frequent *community* clusters in the two sub-corpora.

Comparing the clusters, we can see that the two discourse communities do not share the same contextual collocates for the word *community*. The collocates of *community* for each corpus presented an overview of the word's usage in the two discourse communities, and made it clear that the two communities contextualize the word significantly differently from each other. The sample extended co-texts (7) and (8) of retrieved concordance lines from the *De Beers* annual reports produce a true account of the word's usage.

(7) Wherever we operate, we support countless community needs in education, health and welfare through our corporate social investment programmes. Our motto: give people a hand, not a handout. [...]. Building sustainable futures [...] includes developing local skills [...].

(8) Maintaining good relations with local communities helps ensure our continued access to new supplies, promotes consumer confidence in diamonds and minimises the risk of costly and unnecessary disputes. We see our future as [...] the neighbor of choice to our local communities.

As we can see in excerpts (7) and (8) *community needs* are important intrinsically of themselves, and the diamond industry prioritizes its involvement by investing huge amounts of money into *community* development programmes; but *community* well-being implies a means for acquiring something else in value, such as *access to land, supplies, employees, local skills,* etc. This entails instrumental value. Being a benefactor and having access to assets may be good only to the extent that the 'benefactor' can use them to get something else, when all is said and done, the 'licence to operate'. 'Community-related accountability' discourse is particularly persuasive as *community strategy* rhetoric. In fact, the annual reports to society go to great lengths to create an impression of 'valued communities'. Widely used affirmations are the *community* as *neighbours, partners, togetherness, shared values, collective benefits.* The discursive construction of how the company sees itself in the eyes of its stakeholders, influences the way it builds its relationship with the *local community*, and how the diamond industry responds to conflict and threats to its legitimacy. Moreover, the company frames itself not only as a '*neighbour*', but as the key instigator of economic development and improved standards of living. The rationale behind the existence of a diamond mine is that the community will be better off in both the short and long-term, as diamonds can be "transformed into economic wealth' and improve the "quality of life and wellbeing" (AR 2006). The diamond company positions itself at the 'centre' of *community* life as its *family*, and the community's *lifeline.*

This type of 'life-line' discourse, has been the object of various social science studies (MMSD 2002), which describe the possible negative impacts of the phenomenon of community dependency. No doubt the mining industry has realized huge benefits to local communities improving health and education and so on. However, while the diamond industry (and other multinational mining companies) have 'remodelled' themselves as good corporate citizens, it has been well-documented that in undertaking CSR, the risk may be to create dependency on the company (or on a single commodity), whether intentional or unintentional. In effect, the tone of the narrative appears to be reciprocity, whereby the company gains legitimacy to the land and employees, and the community is given sustained economic development, and social improvement benefits. What is more, the reports show how there is an inclination to define 'community space' with no delineated boundaries or geographic assertions, except in terms of *mining community* delineating the mining area, or the *community* as it exists in the minds of its members, which could be an inclusive term for a diverse range of stakeholders, embracing the local community, employees, traders, etc. These affirmations contrast with the semantic use of *community* in the NGO texts, as shown in the following table:

1	iation by the Southern African Development	**Community**	(SADC), a Government of
2	Cycle of Violence Contributes to Sense of	**Community**	Saves Child Soldiers Personal
3	countries in the KPCS. The international	**community**	should: 1. Ensure that the
4	can an individual help to heal a broken	**community**	so that the community can take
5	upt and corrupting'. 46 This led the Shia	**community**	to transfer their support
6	iamond mine at Gope, a traditional Bushman	**community**	within the reserve. The mine
7	sions and how they impact on the diaspora	**community**	in Africa, particularly

Concordance 8. Sample concordance lines for *community* – Opponents txt.

Community in the sample concordance lines above refers to the *international, diamond, South African, NGO, Bushman, Molapo, Lebanese, Shia, diaspora* communities. The *community* takes on geographical and sociographic space and location, encompassing land which belongs to the named community. They also make an emotional appeal to the audience to understand the *sense of community* (line 2) and the *broken community* (line 4).

To conclude the analysis, two identity themes have emerged in the process of reconstructing corporate reputation: firstly, the company as a world leader and secondly, the company as a motor of economic and social development for sustaining the life of the local community. Taken as a whole, the legitimizing strategy is that of projecting the company's identity, how it views itself, onto the community. Weinstein and Deutschberger (1963) interpret this concept as a conscious strategy of 'altercasting', that is, "projecting an identity, to be assumed by other(s) with whom one is in interaction, which is congruent with one's own goals" (1963: 454).

8. Conclusions

The diachronic and synchronic analyses have shown how identity and reputation construction is an ongoing process over time. This is particularly revealed by the shifting stance the company takes to reposition itself and respond to changing social perceptions and stakeholder interests. We have seen that the company minimizes its risk of losing legitimacy by responsibly reporting on its activities, policies and projects, and by collaborating with other legitimizing actors. So we can refer here to the 'positive' element of the discourse analysis mentioned in the introduction (Martin 2004). However, as the text analysis has shown, there is still a fair amount of ambiguity and contradiction in the discourse. The analysis of corporate legitimacy vis–à-vis the annual reports, which show CSR strategies that focus on tangible projects for attaining 'cognitive' and 'pragmatic' legitimacy, may not be sufficient to evoke 'moral' legitimacy too. The link between legitimacy and responsibility exists, but any explicit dialogicity and interaction between the two agencies, responding directly to each other's arguments, has not been treated in this paper.

However, one of the aims of this chapter has been to provide a framework to study company response to negative events and how these responses restore the company's image and reputation.

The identities acknowledged above emerged from a linguistic analysis that combines corpus linguistic retrieval techniques with the more traditional text-based methods of discourse analysis. The data findings show how the study of collocational properties of clusters, or 'extended units of meaning', and their related pragmatic textual functions, together with semantic prosodies, illuminate the linguistic patterns across the texts or specialized genres, presenting additional perspectives on the understanding of discourse construction. The evidence is that collocation and cluster analyses provide a source of valuable insights on form-function correlations related to phraseology and discourse construction. The chapter has also illustrated how corpus linguistic tools may be used to reveal evaluative meaning in a text focusing on words and phrases, which may not appear to have an explicitly evaluative meaning or be immediately obvious to intuition. The concgram analysis also revealed how lexical items can change meanings together with their associated semantic sequences.

As regards the pedagogical applications, the implications are self-evident. The identification of phraseological patterns across specialized texts, can serve as a tool for textual analysis, and can be used to help raise student awareness of co-occurring words and 'strings of words', which enables students to master the discourses and genres of a specific discipline. On the whole, corpus studies provide insights not only in quantitative terms, but also from a qualitative perspective, as they contribute to the identification of implicit discourse and socio-pragmatic uses of language. The cross-fertilization of linguistics and data driven methodological procedures may potentially shed new light on the patterns that characterize a specialized domain, for example in this case study, in the quest for a better understanding of corporate responses to the reconstruction of reputation and identity.

References

Bakhtin, Mikhail M. 1986. *The Problem of Speech Genres and Other Late Essays.* Austin: University of Texas Press.

Baker, Paul 2006. *Using Corpora in Discourse Analysis.* London: Continuum, Boden.

Berger, Peter / Luckmann, Thomas 1967. *The Social Construction of Reality: A Treatise in the Sociology of Knowledge.* New York: Anchor.

Biber, Douglas 2006. *University Language: A corpus-based study of spoken and written registers.* Amsterdam: John Benjamins.

Bilbow, Grahame 2002. Commissive speech act use in intercultural meetings. *International Review of Applied Linguistics in Language Teaching* 40, 287-308.

Bondi, Marina 2007. Key-words and Emotions: A Case Study of the Bloody Sunday Enquiry. In Fairclough, Norman / Cortese,

Giuseppina / Ardizzone, Patrizia (eds) *Discourse and Contemporary Social Change*. Bern: Peter Lang, 407–32.

Bowker, Janet 2013. Variation across spoken and written genres in internal corporate communication: multimodality and blending in evolving channels. In Bamford, Julia / Diani, Giuliana / Cavalieri, Silvia (eds) *Variation and change in spoken and written discourses: Perspectives from corpus linguistics*. Amsterdam: John Benjamins (forthcoming).

Brown, Andrew / Coupland, Christine 2004. Constructing organizational identities on the web: a case study of Royal Dutch/Shell. *Journal of Management Studies* 41/8, 1325-1347.

Castells, Manuel 2000. *The Rise of the Network Society: The Information Age: Economy, Society and Culture,* Volume I, Oxford: Blackwell.

Cheng, Winnie / Greaves, Chris / Sinclair, John / Warren, Martin 2009. Uncovering the Extent of the Phraseological Tendency: Towards a Systematic Analysis of Concgrams. In *Applied Linguistics* 30/2, 236-252.

Cheng, Winnie / Greaves, Chris / Warren, Martin 2006. From n-gram to skipgram to concgram. *International Journal of Corpus Linguistics* 11/4, 411-433.

Cornelissen, Joep P. / Elving, Wim J. L. 2003. Managing corporate identity: an integrative framework of dimensions and determinants. *Corporate Communications: an International Journal* 8/2, 114-120.

Fairclough, Norman 2003. *Analyzing Discourse: Textual Analysis for Social Research.* London: Routledge.

Firth, John R. 1957. A synopsis of linguistic theory 1930-1955. In Firth, John R. *Studies in Linguistic Analysis*. Oxford: Blackwell, 1-32.

Flowerdew, John 2008. Critical discourse analysis and strategies of resistance. In Bhatia, Vijay Kumar / Flowerdew, John / Jones, Rodney H. (eds) *Advances in Discourse Studies,* London: Routledge, 195-210.

Fombrun, Charles 1996. *Reputation: Realizing Value from the Corporate Image*. Boston, MA: Harvard Business School Press.

Fombrun, Charles / Rindova, Violina 2000. The road to transparency: Reputation management at Royal Dutch/Shell. In Schultz, Majkin / Hatch, Mary Jo / Larsen, Mogens (eds) *The expressive organization, linking identity, reputation, and the corporate brand.* Oxford: Oxford University Press. 77-96.

Garzone, Giuliana / Catenaccio, Paola (eds) 2009. *Identities across Media and Modes: Discursive Perspectives*, Bern: Peter Lang.

Goffman, Eugene 1981. 'Footings'. In Goffman, Erving (ed.) *Forms of talk*. Philadelphia: University of Pennsylvania Press,124-59.

Grieves, Chris 2005. ConcApp 5 software tool. <http://www.edict.com.uk/PUB/concapp/>.

Hardy, Cynthia / Palmer, Ian 1998. *Discourse as a Strategic Resource*. University of Melbourne, Department of Management, 18.

MMSD 2002. Breaking New Ground: Mining, Minerals, and Sustainable Development. *World Business Council for Sustainable Development.*

Hunston, Susan 2002. *Corpora in Applied Linguistics*. Cambridge: Cambridge University Press.

Martin, James 2004. Positive discourse analysis: solidarity and change. *Revista Canaria de Estudios Ingleses* 49, 179-200.

Partington, Alan 1998. *Patterns and meanings*. Amsterdam: John Benjamins.

Partington, Alan 2006. Metaphors, motifs and similes across discourse types: Corpus-Assisted Discourse Studies (CADS) at work. In Stefanowitsch, Anatol / Gries, Stefan Th. (eds) *Corpus-based Approaches to Metaphor and Metonymy*. Berlin: Mouton de Gruyter, 267-304

Salvi, Rita / Turnbull, Judith 2007. The English of Companies online: National Identity and Global Culture. In Bamford, Julia / Salvi, Rita (eds) *Business Discourse: Language at work.* Roma, Aracne Editrice, 9-45.

Schuetz, Janice 1990. Corporate Advocacy as Argumentation. In Trapp, Robert/ Schuetz, Janice (eds) *Perspectives on Argumentation*, Prospect Heights, IL: Waveland Press, 272–284.

Scott, Mike 1999. *WordSmith Tools Help Manual.* Version 3.0. Oxford: Oxford University Press.

Sinclair, John 1996. The search for units of meaning. *Textus* 9/1, 75–106.
Sinclair, John 2004. New evidence, new priorities, new attitudes. In Sinclair, John (ed.) *How to use corpora in language teaching.* Amsterdam: John Benjamins, 271-299.
Spencer-Oatey, Helen 2007. Theories of identity and the analysis of face. *Journal of Pragmatics* 39/4, 639-656.
Suchman, Mark C. 1995. Managing Legitimacy: Strategic and Institutional Approaches. *The Academy of Management Review* 20/3, 571-610.
Tognini-Bonelli, Elena 2001. *Corpus Linguistics at Work.* Amsterdam: John Benjamins.
van Dijk, A. Teun 2001. Critical Discourse Analysis. In Schiffrin, Deborah / Tannen, Deborah / Hamilton, Heidi (eds) *The Handbook of Discourse Analysis.* Oxford/Malden, MA: Blackwell.
van Dijk, A. Teun 2005. Contextual knowledge management in discourse production. In Wodak, Ruth / Chilton, Paul (eds) *A New Agenda in (Critical) Discourse Analysis.* Amsterdam: John Benjamins, 71-100.
Weinstein, Eugene / Deutschberger, Paul 1963. Some dimensions of altercasting. *Sociometry. Journal of Interpersonal Relations* 26, 454-466.

DONATELLA MALAVASI

'Made in Italy': Local and Global Distinctive Traits

1. Introduction

Over the last few decades, communication and business communication have been performed using "[n]ew communication tools and channels [...] that act as complements or substitutes for the way communication has been performed in the past" (Deller/Stubenrath/Weber 1999: 351). There is little doubt indeed that the Internet has become the preferred conduit for disseminating information, distributing companies' corporate communications, and undertaking a range of activities aimed at disclosing data efficiently and at a low cost. Following the inception of the World Wide Web in the mid-1990s, companies have begun to establish their own corporate website as a platform to present and spread data. Originally, the need for a website was not fully understood by those companies, whose presence over the Internet has been motivated primarily by imitation or the so-called 'me too' factor (Perry/Bodkin 2000; Gowthorpe 2004). On the contrary, a high number of firms have realized that some type of web presence is necessary and convenient to gain significant marketing advantages.

In the light of recent research on the subject (Boardman 2005; Pollach 2005; Bamford/Salvi 2007; Garzone/Poncini/Catenaccio 2007; Isaksson/Jørgensen 2010), websites appear to prioritize 'product' and 'performance' related information, and to incorporate "a mix of promotional elements, such as press releases, advertisements, games, free gifts and pricing information" (Perry/Bodkin 2000: 87). Apart from their unchallenged and undisputed informative nature, website activities involve much advertising and (self-)promotion (Lymer 1999). Notably, websites are recognized to act as a virtual shop window, having the overt intent to market a product/service, attract clients, influence their behaviours, promote brand identification, gain support of the firm's internal staff, and enhance investors' trust. In detail, company websites are to be seen as corporate communication forms which are concerned with "establishing and maintaining long-term relations with those audiences that are perceived by an organisation as being those that help to produce successful outcomes for corporate strategy" (Dolphin 2004: 27). It is also widely accepted that corporate communication practices, including corporate websites, perform the function of promoting

> a strong corporate culture; a coherent corporate identity; a genuine sense of corporate citizenship; an appropriate and professional relationship with the press; a quick and responsible way of communicating in a crisis or emergency situation; an understanding of communication tools and technologies; a sophisticated approach to global communication. (Goodman 2000: 70)

Indeed, a presence online allows companies not simply to disclose cheaply and timely multimodal information about themselves and their products/services, but also to achieve something more solid, durable and long-term, i.e. the construction of a sound corporate identity and the projection of a commendable image. Specifically, on the one hand, corporate image corresponds to the firm's portrait made in the mind of a consumer, whereas, on the other, corporate identity has to be understood as "the articulation of what an organization is, what it stands for, what it does and how it goes about its business [...]" (British Standards Institution, 1995), (Topalian 2003: 1119; see also Bettetini 1993; van Riel/Balmer 1997; Balmer 1998 and 2001; Nguyen/Leblanc 2001; Melewar/Karaosmanoglu 2006). What inspires and forges company identities is the set of cultural values that unite an organization around a credo, a range of timeless guiding principles, an admirable vision, a unique mission, history and philosophy. Thus, in an attempt to build an attractive identity, companies appear to showcase their best attributes, 'advertise' their distinctive traits to an international audience, and give illustrations of their expertise, dependability and reliability. Echoing Isaksson and Jørgensen's model (2010), firms turn out to rely on the web to communicate their credibility by highlighting their ethos qualities of expertise, trustworthiness (i.e. integrity, truthfulness, courage and passion), and empathy (i.e. attention to clients' needs and clients' enjoyment derived from dealing with the company). Therefore, for the enactment of a consistent identity it can be easily assumed that in their websites companies insist on the values which they consider fundamental to reaching a distinctive position in the wide economic scenario, overtaking competitors, and nurturing commitment among stakeholders.

Resting upon the previous general definitions, this chapter sets out to examine a sample of corporate websites and to investigate how identity is enacted by a selection of Italian firms, which work for the three key sectors of 'Made in Italy', namely, Fashion, Food and Furniture. Specifically, identity will be examined in its discursive construction or, to put it in other terms, as a phenomenon which is actively, ongoingly, dynamically constituted in discourse (Benwell/Stokoe 2006: 4). Among the strategic factors and discourse resources which contribute to a laudatory self-portrayal of companies, the analysis will focus on the interconnectedness of space, identity and discourse. Particular attention will be paid to the representation of space in the website of the firms under study and the role it takes up in their self-appraisal.

The materials used for the study as well as the methodology adopted will be described in the next section. This will be followed by a quantitative as well as qualitative analysis (Section 3), and by a discussion of the main findings obtained (Section 4).

2. Materials and methods

The study is based on a small corpus which is made up of twelve websites (approximately 150,000 running words) belonging to some renowned companies whose products have become symbols of the so-called Made in Italy. The choice of websites in their English version focused on the '3Fs', that is to say, the three main sectors which are commonly associated with Made in Italy, namely Fashion, Food and Furniture (Decree No. 205 issued on the 10th July 2008 by the Italian

Ministry of Economic Development; Bucci/Codeluppi/Ferraresi 2011; Secchi 2012). For each sector four leading groups were identified and selected as good representatives of the well-known 'Italian' brand, quality, style and tradition (Benetton Group, Marzotto Group, Max Mara Fashion Group and Miroglio Group for Fashion; Barilla, Gruppo Cremonini, Ferrero and Parmalat for Food; Marazzi, Gruppo Mauro Saviola, Natuzzi and Molteni for Furniture).

As regards the criteria for corpus design, only the verbal text of specific sections of each website was taken into account, while financial data, news, press releases, contact details, tables, images, videos, and many other multimodal resources were not looked at. In detail, the sections of the websites which constitute the corpus are the manifold web pages which come under a number of headings and sub-headings such as *About us, The Group, Company Vision, The History, Brands, Products, Sustainability,* to name but a few. The choice of these sections was motivated by the very nature of these portions of websites, whose primary aim is the dissemination of corporate information. These segments of websites, which serve as information provider tools, are deployed by each company to offer Internet surfers accurate and user-friendly information about its businesses, and, concomitantly, to encourage the audience to perceive it as a trustworthy rather than deceitful entity.

From a methodological perspective, the study was carried out with the support of corpus linguistics tools such as wordlists, KWIC concordances and collocates (Sinclair 1991, 1996, 2003, 2004).[1] In an effort to identify the role that the notion of space and its local vs. global dichotomy play in the shaping of corporate identities, the frequency wordlist was generated and inspected for the purpose of deriving a sample of relevant space-related items. The quantitative and qualitative study of some markers of space was meant to explore the strategies developed by firms to build their identity around their national origins, on the one hand, and their international vocation and global appeal, on the other.

---

1 For the processing of wordlists, concordances and collocates, the PC software *Wordsmith Tools 5.0* (Scott 2008) was used.

A selection of lexical indications of space was concordanced and analysed in the extended lexical units in which they occur as well as collocational patterns (Sinclair 1996, 2003, 2004). The occurrences of the nodes in longer stretches of language were examined in an attempt to recognize the 'tactical arsenal' of resources harnessed by companies to index and give prominence to their global and local gist or essence.

3. Results

As anticipated in Section 2, this section is intended to illustrate the main findings of the analysis of some space-related lexical signposts or signals pointing to companies' self-representation as locally and globally proactive agents. By applying the criteria formulated in the previous section, the list of the most frequent words was processed and a sample of lexical items (reported in Table 1 below) was carefully chosen among the 200 most widely spread tokens of the corpus by virtue of their hypothesised contribution to the formulation and enactment of corporate identity in space.

N	*WORD*	*FREQ.*	*%*
47	WORLD	232	0.16
67	INTERNATIONAL	184	0.12
71	ITALY	181	0.12
92	ITALIAN	144	0.10
171	LOCAL	94	0.06
174	GLOBAL	92	0.06

Table 1. Space formulations and their occurrences.

The representation of space and its topicalisation in websites will be accounted for in its twofold dimension: local/national and international. The analysis of firms' national distinctive traits (Sub-section 3.1.) will be followed by the study of companies' global reach and international flavour (Sub-section 3.2.). A combination of patterns will provide some evidence of the strategies developed by the Italian groups to promote themselves as dependable actors that rely on their spatial specificity while consolidating their premium reputation worldwide (Section 4).

3.1. Italianness

The focus of this section is on the analysis of some lexical items which refer to the local and national characterisation of the companies under study. It can be assumed indeed that the self-representation of these companies, which are icons of Made in Italy, revolves around their Italian roots and locatedness. Spatial formulations of firms' origin and 'domestic' specificity are *Italy, Italian* and *local.*

The inspection of the concordances of the search word *Italy* (181 entries) suggests that it is recurrently used to refer to and specify the geographical location of each company's headquarters, production plants, trading offices and points of sale. Specifically, the co-occurrence of *Italy* with precise place and town names (e.g. *Santeramo in Colle, Parma* and *Castrette* in examples 1-3) can be interpreted as firms' attempts to situate the readership within their homeland. The continual flagging or reminding of nationality is testified to by excerpts 1 to 3 (my italics as in the other quotations), which show that considerable emphasis is placed on the spatial context in which firms and their production sites are located.

(1) Pasquale Natuzzi personally coordinates this strategic activity from the *Santeramo in Colle, (Italy)* Style Centre where a team of 120 professional designers, architects, colour experts, craftsmen, engineers and furniture specialists is devoted entirely to design quality, studying trends and choosing materials. (furniture\natuzzi)

(2) Barilla Group. Barilla, originally established in 1877 as a bread and pasta shop in *Parma, Italy,* ranks as one of today's top Italian food groups. (food\barilla)

(3) Logistics Benetton has direct control of the logistics phase for both own manufactured and sourced products, and has invested in modelling, organization, and automation of logistic processes in order to completely integrate the entire production cycle, from client orders, to packing and delivery. Automated Sorting System The state-of-the-art logistics operation at *Castrette (Italy)* has a fully automated innovative sorting system, whose propulsion is based on electromagnetic fields, capable of handling individual orders for over 6,400 Benetton shops worldwide. (fashion\benetton)

However, by repeatedly referring to *Italy,* companies endeavour not only to signal their 'physical' and 'material' presence in this country but also to stress the more 'intangible' relation they have established with the territory and its culture. Accordingly, irrespective of the industry they work for, the integration and interconnectedness of the firms and the geographical area in which they operate manifest themselves in a twofold manner. On the one hand, companies are depicted as 'attentive' entities who conform to the Italian cultural and legal specificities or the *best practices from Italy* (example 4). On the other, these firms profile themselves as proactive agents who contribute to the preservation and proliferation of Made in Italy products in all their distinctive marks – creativity, style, glamour and quality (examples 5 and 6).

(4) By way of emulating an entrepreneurial tradition that sits at the heart of a family, the parent company clearly states in this booklet the ethical business principles that reflect and reinforce who we are, our heritage, our purpose and values. This booklet has evolved from the Miroglio Group Code of Ethics and *takes into account global standards and legislation, as well as cultural and legal best practices from Italy* and from the local markets in which we operate. (fashion\miroglio)

(5) The Max Mara Group treasures and constantly develops its know-how regarding the manufacture of garments. For this reason a company of the Group is *dedicated to representing the excellence of Made in Italy products* as regards the manufacture of women's coats and jackets. (fashion\maxmara)

(6) Today, a capacity for innovation is an essential competitive asset. But one thing has remained the same: quality, the common thread that for four generations has distinguished the evolution of a group that is one of the world leaders in the furnishing and design sector, *a testament to made in Italy products* in over 60 countries. (furniture\molteni)

Thus, Italianness, which is to be intended in its 'tangible' and spatial sense but also metonymically as a synonym of trustworthiness, veracity and experience, is given prominence to in the selected websites and becomes a key component of firms' identity. Because of their being Italian, companies appear to be characterised by part of the values which are commonly associated with their homeland and Made in Italy as well. Therefore, these entities' origin, which is insisted on as an extra guarantee of reliability, is to be interpreted as a relevant constituent of their self-representation as repositories of 'national values' such as quality, tradition, and *savoir-faire*.

Some of the features with which companies and their self-portrait are imbued by their nationality can be additionally recognised by skimming the concordances of the search word *Italian* (144 entries). In an attempt to captivate the audience, inspire confidence and create goodwill with the public, the groups under consideration appear to emphasise some of the values which are 'labelled' as peculiar to their country of origin. Evidence from the corpus suggests that the tenets which Made in Italy hinges upon and which come to be an integral part of Italian companies' value system are *style, taste* and *creativity* (7), *excellence* (8), *expertise* (9), *flair, know-how* (10), *quality,* and *tradition.*

(7) IDEA The line of printed fabrics realized in China and imbued end-to-end by *the style and taste conceived entirely in Italy.* Idea looks to the Chinese home market and to the major international customers that produce the finished garment in the Far East. Speed, superb value for money and *Italian creativity* are sublime add-ons for this collection. (fashion\miroglio)

(8) *Italian Food Excellence* We are The Italian Food Company. Since 1877 - *the Ambassador of Italian food culture in the world.* Our products and brands are deeply anchored in our culinary roots, and our aspiration is to become the leading authority and center of *excellence of Italian culinary art and food.* Knowledge & Innovation Barilla aspires to be home of competence and hub for innovation to nurture *Italian food as a longlasting concept.* (food\barilla)

(9) The Group currently boasts manufacturing plants in Italy, Spain, France, Russia and the United States, many commercial branches and showrooms providing work for over 6,000 employees. From its own production sites the Group exports to the entire world, augmenting *Italian expertise* with the

lifestyles and requirements found in the various local markets. (furniture\marazzi)

(10) Barilla has become one of the world's most esteemed food companies and is recognized worldwide as a symbol of *Italian know-how* by respecting its longstanding traditional principles and values, considering employees a fundamental asset and developing leading-edge production systems. (food\barilla)

The repertoire of positive features which are given prominence to in the manifold references to Italy and Made in Italy is strategically and wisely resorted to by the selected companies to underline their expertise, and to reassure both existing and prospective customers about the quality of the products they manufacture. For the establishment of an attractive identity, Italian firms appear to portray themselves as symbols and ambassadors of those prerogatives or peculiarities which derive from their local, regional and national roots. Thus, more or less overt expressions of dependability, skills, competence and traditions accumulated over time and handed down from generation to generation are interwoven with rhetorical devices which foreground the authenticity, veracity, genuineness and naturalness of the companies' products and production processes. Specifically, production methods, procedures, and ingredients or raw materials are presented in their geographical 'limitedness' as traditionally Italian, hence rigorous, high-quality and genuine. Typical examples are:

(11) CUOR DI PAESE ORO is *a complete Italian product*, in fact *it is produced only using selected meats originating from Italian breeding*. Cuor di Paese Oro adds to its excellent taste, typical of Cuor di Paese mark, the *certainty of the entire traceability*. (food\cremonini)

(12) We spend 700,000 hours per year focusing on quality. We design and study, photograph, assemble and describe. Since 1934 *the entire Italian chain of production has been centred in Giussano, genuine Made in Italy*. (furniture\molteni)

(13) The quality of the product comes from the great care taken in choosing raw materials as well as the lengthy experience in processing natural fibers such as fine merino wools, precious cashmere and other sophisticated fabrics. The

design and research *team brings together structure and materials with Italian quality and taste in a harmony of colors and shapes.* (fashion\marzotto)

The authenticity of 'home-made' products is exalted and celebrated in combination with companies' respect for traditional production methods, severe quality controls and careful selection of the local and 'best ingredients'. In addition to the concerted efforts made by each firm's management to create products that are safe and easily accessible, another substantial determinant of companies' self-appraisal is the pervasive insistence on their CSR (Corporate Social Responsibility) activities and their commitment to sustainability.

In detail, the examination of the occurrences of *Italy* and *Italian* reveals that the firms under study strive to parade an ethos of legitimacy and reliability also by referring to their contribution to cultural and social events, engagement in philanthropic projects, support for the local community, and commitment to the safeguard of the environment (examples 14 to 18).

(14) To achieve its objectives Barilla always... Operates respecting the rights of present and future generations, communities, People and suppliers. Creates opportunities for the areas where it operates and their People, *promotes local development and culture in Italy* and worldwide. (food\barilla)

(15) Sport teaches us how to behave in a social context, to be honest when faced with problems, to accept success and failure in equal measure. It is for this very reason that Kinder has been *collaborating with and supporting the main Italian* and European *sports federations and institutions for many years*, in order to develop activities aimed at creating a generation that grows up well and truly 'on the move'. (food\ferrero)

(16) Benetton Group is committed to being a globally responsible company in social, environmental and economic terms: this is a commitment for present and future generations, *to grow together with the community in which the company operates, in Italy* and all other markets. (fashion\benetton)

(17) Revision of the distribution process, such as, for example, *the "Courier" project in Italy* with small to medium-sized local carriers which, through the optimization of delivery and order taking routes for small retailers, of which there is a particularly great proliferation in Italy, has enabled a 17% reduction in the kilometers traveled per year by individual carriers. (food\ferrero)

(18) Ecological Panel Qualitative controls on the incoming wood Every year the companies of the Ecological Panel *recover 1,5 millions of tons of waste wood in Italy* and Europe, thanks to a network of 30 collection centres. The collection is mostly by rail to reduce the impact of the transports on the environment. Strict quality controls are made on the wood to guarantee chemical and physic qualifications in observance to the highest security standards. (furniture\maurosaviola)

In their web-based self-descriptions, companies tend to stress their responsible orientation towards sustainable economic development, while placing emphasis on the nature of their undertakings as people-centred. The CSR-oriented nature of Italian firms' identity emerges strongly from the passages of websites in which *local* occurs. The concordances of *local* (94 occurrences) demonstrate that it is on this node that firms centre convincing statements (as in the following examples) about their benevolence, and the multifarious social and philanthropic activities they sponsor to support communities.

(19) The Parmalat Group is centrally coordinating its operations in various regions of the world *to develop a range of functional products that are sufficiently diversified to meet the needs of local populations.* In the West, the focus will be on light functional products, while in the developing countries, the Group will be creating and testing the feasibility of projects to deliver proper nutrition (vitamins, proteins, calcium, sugar). (food\parmalat)

(20) 3.7 HEALTH AND SAFETY We recognise and affirm the importance of health and safety within our business. We seek *to provide a healthy, safe and clean working environment in line with local laws, regulations and industrial practice.* We measure, appraise and report performance, as part of our commitment to the health and safety of our employees, contractors and everyone who works on or visits our sites. (fashion\miroglio)

(21) It should be noted that the Ferrero Foundation is not a "Granting Foundation", that is, it does not provide loans or contributions to other organizations or to social, cultural or humanitarian institutions. Rather, is an "Acting Foundation", which means it develops, on its own, *various social and cultural initiatives (the latter specifically related to the local Alba community or Piedmont region*). (food\ferrero)

(22) [...] the group considers working to guarantee social and environmental sustainability a core value of its company strategy: [...] in its impact on communities - *To prevent potential negative impacts on local communities in*

> *which it operates*, the Benetton Group is progressively applying a human rights impact assessment tool in collaboration with the Valore Sociale organisation, a CSR centre of excellence whose members include NGOs, associations and universities. (fashion\benetton)

Repeated mentions of the companies' attempts to provide their proactive contribution to customers' satisfaction (example 19) and the protection of the environment at large are closely intermingled with declarations of firms' whole-hearted support for the local communities' rights, cultural development and well-being (examples 21 and 22). Accordingly, the construction of an enduring corporate identity is suggested to rest on companies' self-presentation as responsible citizens who strive to impact positively on local and regional populations. Signals pointing to companies' orientation towards local communities, safeguard of the environment and adherence to national standards, principles and regulations (see example 20) turn out to be key strategies adopted by these economic actors to increase their credibility, and enhance perception of legitimacy and trustworthiness.

As a result, the analysis of the markers of space *Italy, Italian* and *local* shows that the basis for the companies' enactment of a favourable identity is the insistence on all those pre-eminent features which are inextricably linked to their homeland. To arouse trust and esteem towards themselves in the general public, Made in Italy companies have been found to stress their national (but also global, as will be demonstrated in sub-section 3.2.) prestige, specificity, uniqueness, and deeply-rooted tradition. In this sense, an array of space-related values has been suggested to underlie firms' self-promotion as actors committed to the highest national standards of quality, adherence to traditional production processes, community activism, and preservation of all those values which Made in Italy symbolises and embodies, that is creativity, style, glamour and excellence.

3.2. Internationality

In this sub-section, the scope of the study will be extended to account for another range of spatial formulations which are still assumed to be a key component of corporate identity albeit they index the international rather than national dimension of space. Among the lexical items which mark the worldwide side of space, particular attention will be paid to the following emblematic nodes: *world* (232 entries), *international* (184 occurrences) and *global* (92 hits).

The analysis presented in the previous sub-section reveals that the companies' pride in their national origin and specificities is commonly accompanied and counterbalanced by a more international vocation. The inspection of the concordances of *Italy, Italian* and *local* show that these search words cluster with signals pointing to companies' global essence and coverage. The see-sawing between firms' strong ties with the local and national territory and their international flavour is testified to by mentions of national locatedness along with references to the wider economic scenario, as in the following examples.

(23) After more than fifty years of challenges and intuitions, the Natuzzi Group is the top Italian producer for interior design, with 6 factories operating *in Italy and* 5 *around the world* (China, Brazil and Romania), turnover of 486.4 million Euros (as of 31st December 2011) and 6,616 employees. (furniture\natuzzi)

(24) Barilla owns 43 production sites (13 *in Italy and* 30 *outside Italy*), of which 9 are directly managed mills that provide most of the raw materials for the production of its pasta and bakery products. Barilla exports to more than 100 countries. (food\barilla)

(25) Today, the Cremonini Group is one of the biggest industrial food players in *in Italy and Europe*, characterized by a managerial approach to the market with strong growth prospects. (food\cremonini)

(26) Vision Benetton Group is committed to being a globally responsible company in social, environmental and economic terms: this is a commitment for present and future generations, to grow together with the community in which the company operates, *in Italy and all other markets*. (fashion\benetton)

(27) Marazzi Group is the leading player in the ceramic sector in many of the most important international markets: number one *in Italy, France and Russia*, number two in the USA and one of the top contenders in Spain, Germany, Greece and Scandinavia. This widespread and extensive geographic coverage undoubtedly represents a distinctive characteristic of the Group. (furniture\marazzi)

The combination of markers of local and global space draws attention to the national and international relevance of companies' competitiveness, leadership, social commitment and profitability. In detail, firms' capability to take a leading role in the global market can be perceived if the occurrences of the node *world* (232 entries) are investigated. The examination of the search word suggests that it occurs in many stretches of language (see the following exemplifications) which share a common emphasis on firms' global presence and prestige.

(28) The Molteni Group has a global presence, distribution in over 60 countries *around the world*, 5 Flagship Stores (Milan, Paris, London, New York and Singapore), over 30 single brand stores and 15 executive offices in Italy and abroad as well as over 600 stores. (furniture\molteni)

(29) For 175 years the Marzotto Group has stood for competitiveness and innovation. A *world leader* in the textile industry, it is highly visible from yarns to fabrics. The Marzotto Group operates in the textile sector with prestigious brands that make it a major player in terms of quality, quantity, R&D on *world* fabric and yarn *markets*. (fashion\marzotto)

(30) […] in Japan, Brazil and Poland we already have active production plants and in other countries we have come to a joint venture and collaboration agreement with local companies. We export our products to 22 countries *around the world*. (food\cremonini)

Recursive references to the location of stores and factories all around the world add their contribution to companies' self-appraisal as actors whose prominent position is visible and recognisable worldwide. Evidence in support of the global role that Made in Italy companies ascribe to themselves is brought by repeated expressions which highlight the internationality of their clientele. By insisting on their global and diversified consumers, firms appear to prioritize the appreciation and consumption that their brands and products are granted by customers from every corner of the globe (examples 31 and 32).

(31) The brands under which the Group markets its products enjoy *extraordinary name recognition throughout the world* among designers, architects and construction firms, as well as high positioning and genuine admiration on the part of the vast legions of retail customers. (furniture\marazzi)

(32) Since then, Nutella continues to *inspire and satisfy the palate of the entire world*: from the United States to Europe, Australia to Canada, its unique taste and its distinctive jar appeal to consumers of all generations. (food\ferrero)

Therefore, the highly-reputed nature of these firms is reinforced by positive evaluations of their abilities to satisfy clients' needs, and to respond to people's growing concerns about the quality of life 'here and in the rest of the world'. Accordingly, Italian groups' corporate websites are pervaded by detailed tracts full of claims about their involvement in important international projects (example 33), and CSR-oriented ambitious initiatives promoted to support communities and *consumers throughout the world* (34).

(33) Kitchens sky high. Dada furnishes the Strata Tower, London's tallest building and Burj Khalifa, the *world's tallest skyscraper*. (furniture\molteni)

(34) The Parmalat Group is an Italian food group with a multinational strategy that seeks *to increase the well-being of consumers throughout the world*. The ultimate purpose of the Group is to create value for its shareholders while adhering to ethical principles of business conduct, to perform a useful social function by fostering the professional development of its employees and associates, and to serve the communities in which it operates by contributing to their economic and social progress. (food\parmalat)

A similar contribution to companies' self-description is provided by the nodes *international* (184 occurrences) and *global* (92 hits). Although both adjectives appear in constructions which further insist on firms' presence and recognition outside their national boundaries (example 35), *international* tends to prevail in self-promotional statements which 'advertise' the prizes, acclaim, applauses and certifications that companies have been awarded (examples 36 and 37).

(35) Today many Ferrero *products are "global"*, in that they are sold everywhere. Ferrero is now the fourth largest confectionery group in the world. However Ferrero's distinguishing characteristic is that it is "glocal" (think globally, act locally), a *company which is both global and local*, which focuses on international development, without losing sight of its relationship with local communities. (food\ferrero)

(36) ISO 14001 During 2004, the Marazzi Group obtained ISO 14001 *international certification* which sets the prerequisites for the implementation, development and maintenance of an environmental management system, undertaking to control the environmental impacts of its production process with a view to continual improvement. (furniture\marazzi)

(37) 2009 Ferrero wins the prestigious *international prize*: the "Reputation Award". (food\ferrero)

As testified by the previous examples, a crucial role in the accomplishment of self-appraising tasks is played by some objectifying techniques (see references to prizes or certifications), which are utilised to provide third-party evidence and to convey (quasi-)'factual' information. Furthermore, the occurrences of the node *international* reveal that credibility is also nurtured by the considerable emphasis placed on firms' compliance with both national and *international laws, regulations, treatises* (example 38), legislation and practices.

(38) The company, in particular, applies *international laws, regulations and treatises* concerning the employment of women and children, contracts, discrimination, personal health and safety, fair wage conditions and respect for the environment. (fashion\benetton)

In light of the considerations put forth so far, not only Italianness but also internationality appears to be a key driver of firms' identity construction. In this sense, a laudatory self-presentation is to be traced back to companies' insistence on their global presence, popularity, high reputation worldwide, international profitability, and dedication to CSR-related initiatives in many different countries.

4. Conclusion

The quantitative and qualitative analysis of a selection of space-related search words has revealed that companies' self-portrayal relies on two complementary sets of elements. On the one hand, firms turn out to profile themselves as the repositories of those values which are considered specific to Italy and Made in Italy, in particular quality, tradition, expertise and style. On the other hand, Made in Italy groups have been seen to depict themselves as economic actors who are appreciated worldwide, have reached a global market and strive to consolidate their international competitiveness.

This dual interpretation of spatial locatedness as 'within and outside domestic boundaries' underlies the twofold identity of companies: national and international. The specificities of their homeland, the strong ties they have with the local, regional, and national territory, the thriving activities they perform *in loco* are clear marks of the importance that Made in Italy companies attribute to their Italianness. Their national origin and rootedness, which become synonyms of authenticity, veracity, experience and tradition, are leveraged on by the firms under study to establish credentials and inspire a wealth of favourable responses in the audience. Moreover, additional support, trust and confidence are elicited by these firms thanks to the emphasis they place on the prestige they have acquired worldwide. In an attempt to attract and retain clients, inspire confidence among stakeholders, and foster goodwill and approval, companies highlight their international competitiveness and profitability, their respect for foreign markets' rules and cultures, and the CSR activities which benefit an international audience.

Therefore, the study presented in this chapter further confirms that identity is a discursive accomplishment: who we are to each other, then, is not a merely reflexive cognitive phenomenon but rather a practice which is managed, negotiated and produced in what, in this particular study, consists in a treatment of written texts. Specifically, some light is shed on identity as "the situated outcome of a rhetorical and interpretive process in which interactants [i.e. corporate actors in this case] make situationally motivated selections from socially constituted repertoires of identificational and affiliational resources and craft these semiotic resources into identity claims for presentation to others" (Bauman 2000: 1).

In this respect, among the several stylistic, grammatical and lexical frameworks which are commonly accepted to contribute to the construction of identity (e.g. the manipulation of agency, pronominal usage, mood choices, modality, figurative and metaphorical use of language, ideologically biased presuppositions and assumptions, see Fairclough 1992; Benwell/Stokoe 2006), formulations of space have been suggested to take up a key role in the web-based construction and projection of a firm's self.

Obviously, the spatial formulations used in corporate websites are not limited to the six search words analysed in the chapter, as reported in Table 1. Nonetheless, the corpus-based findings documented here reasonably suggest that space and spatial elements are key components of corporate identity. It can be hoped that further research will provide more substance to the inextricable links existing between space, identity and discourse.

References

Balmer, John M. T. 1998. Corporate Identity and the Advent of Corporate Marketing. *Journal of Marketing Management* 14/8, 963-996.

Balmer, John M. T. 2001. Corporate identity, corporate branding and corporate marketing. Seeing through the fog. *European Journal of Marketing* 35/3-4, 248-291.

Bamford, Julia / Salvi, Rita (eds) 2007. *Business Discourse: Language at Work*. Roma: Aracne.

Bauman, Richard 2000. Language, Identity, Performance. *Pragmatics* 10, 1-5.

Benwell, Bethan / Stokoe, Elizabeth 2006. *Discourse and Identity*. Edinburgh: Edinburgh University Press.

Bettetini, Gianfranco 1993. *Semiotica della Comunicazione d'Impresa.* Milano: Studi Bompiani.
Boardman, Mark 2005. *The Language of Websites.* London: Routledge.
Bucci, Ampelio / Codeluppi, Vanni / Ferraresi, Mauro (eds) 2011. *Il Made in Italy.* Roma: Carocci.
Deller, Dominic / Stubenrath, Michael / Weber, Christoph 1999. A survey on the use of the Internet for investor relations in the USA, the UK and Germany. *The European Accounting Review* 8/2, 351-364.
Dolphin, Richard R. 2004. The strategic role of investor relations. *Corporate Communications: An International Journal* 9/1, 25-42.
Fairclough, Norman 1992. *Discourse and Social Change*. Cambridge: Polity Press.
Garzone, Giuliana / Poncini, Gina / Catenaccio, Paola (eds) 2007. *Multimodality in Corporate Communication. Web genres and discursive identity*. Milano: Franco Angeli.
Goodman, Michael B. 2000. Corporate Communication: the American Picture. *Corporate Communications: An International Journal* 5/2, 69-74.
Gowthorpe, Catherine 2004. Asymmetrical dialogue? Corporate financial reporting via the Internet. *Corporate Communications: An International Journal* 9/4, 283-293.
Isaksson, Maria / Jørgensen, Poul E. F. 2010. Communicating corporate ethos on the Web. The self-presentation of PR agencies. *Journal of Business Communication* 47/2, 119-140.
Lymer, Andrew 1999. The Internet and the future of corporate reporting in Europe. *The European Accounting Review* 8/2, 289-301.
Melewar, Tengku C. / Karaosmanoglu, Elif 2006. Seven dimensions of corporate identity. A categorization from the practitioners' perspectives. *European Journal of Marketing* 40/7-8, 846-869.
Nguyen, Nha / Leblanc, Gaston 2001. Corporate image and corporate reputation in customers' retention decisions in services. *Journal of Retailing and Consumer Services* 8/4, 227-236.

Perry, Monica / Bodkin, Charles 2000. Content analysis of *Fortune* 100 company Web sites. *Corporate Communications: An International Journal* 5/2, 87-96.

Pollach, Irene 2005. Corporate self-presentation on the WWW. Strategies for enhancing usability, credibility and utility. *Corporate Communications: An International Journal* 10/4, 285-301.

Riel, Cees B.M. van / Balmer, John M. T. 1997. Corporate identity: the concept, its measurement and management. *European Journal of Marketing* 31/5-6, 340-355.

Scott, Mike 2008. *WordSmith Tools (Computer Software. Version 5.0).* Liverpool: Lexical Analysis Software.

Secchi, Raffaele (ed.) 2012. *Supply Chain Management e Made in Italy. Lezioni da nove casi di eccellenza.* Milano: Egea.

Sinclair, John M. 1991. *Corpus Concordance Collocation.* Oxford: Oxford University Press.

Sinclair, John M. 1996. The search for units of meaning. *Textus* 9/1, 75–106.

Sinclair, John M. 2003. *Reading Concordances*. Longman: London.

Sinclair, John M. 2004. *Trust the Text. Language, Corpus and Discourse*. Routledge: London.

Topalian, Alan 2003. Executive perspective: 1. Experienced reality. The development of corporate identity in the digital era. *European Journal of Marketing* 37/7-8, 1119-1132.

Websites

Fashion

<http://www.benettongroup.com/>
<http://www.marzotto.it/e_il_gruppo.html>
<http://www.maxmarafashiongroup.com/en/home-page>
<http://mirogliogroup.it/en/>
(last accessed in September, 2012)

Food

<http://www.barillagroup.com/>
<http://www.cremonini.it/en/home/>
<http://www.ferrero.com/>
<http://www.parmalat.com/en/>
(last accessed in September, 2012)

Furniture

<http://www.marazzi.it/en>
<http://www.grupposaviola.com/en/>
<http://int.natuzzi.com/>
<http://www.molteni.it/>
(last accessed in September, 2012)

GILLIAN MANSFIELD / FRANCA POPPI[1]

Selling Who We are and /or What Our Product is: Self-presentational Strategies in Italian Company Websites and Product Promotion

1. Introduction

With the advent of globalization, companies have started to feel the overriding need to present themselves not exclusively to the local marketplace, but to the global world as well. Accordingly, if on the one hand the intensified flow of capital, goods and technology has spurred companies to focus on expressing their uniqueness, on the other it has made them aware of the importance of being part of a community. This is why the recently coined term 'glocalization' is becoming more and more widely used, since it focuses on the interplay of the two interdependent concepts of globalization and localization.

In fact, well-known Italian brands of products, distributed on a worldwide scale, are often recognised for the country-of-origin effect they create in global advertising (Melewar/Karaosmanoglu 2006: 856). In comparison, an essential domestic message, displayed either on a company website or in the paper or audiovisual advertisement of the product itself, is likely to undergo some modification and/or adaptation when re-written or dubbed for the global market. The reason for this is not simply to satisfy the marketing style of the country but more importantly, to project a clearly identifiable and relevant cultural self to the foreign market. In the latter case, redundancy in meaning (identifiable from a multimodal analysis) may be a useful factor in re-

1 The authors have jointly developed the whole topic and sections § 1, 5, 6, 7 and 8. Moreover, Gillian Mansfield is responsible for sections § 2 and 3 and Franca Poppi is responsible for section § 4.

inforcing the identity of the manufacturers and/ or their products and country of origin.

Maintaining national identity in a global market has special implications for the linguistic fetishes (Kelly-Holmes 2005) created around a product that the national manufacturer not only desires to project by maintaining essential lexis in the original language (foreignisation strategies), but also in expressions of self-glorification (e.g. "Italy's No. 1 Pasta, Britain's best…") as well as personal commitment to the achievement of perfection in the final product. In doing so the company manages also to project its image into a wider market by focussing on the individual country ("The world's best, the world's number one…"), thus extending the relevance of space further afield.

By means of a series of qualitative analyses, the present chapter purports to reveal how companies in totally different sectors deploy strategies aimed at selling themselves and their product(s) not only to the local markets, but also to the global world. Attention will focus equally on seeking out keywords and their connotative features and on product denomination, by integrating a corpus perspective with interpretative insights from discourse analysis in relation to the construction of identity and to the role of the perception of space in this process, which can imply both foreignising and domesticating tendencies. Results will show that products from two extreme ends of the market sector (from everyday consumer and luxury products to more specialized commodities such as Italian ceramic tiles), nevertheless have a common denominator in terms of the fact that they both contribute to establishing and reinforcing identity in relation to a sense of place and belonging to a particular national and/or social group. Furthermore, this analysis will also highlight the underlying values conveyed implicitly in the promotional message, be it on a website or in the individual product advertisement.

2. The issue of culture

Van Raaij (1997: 261) claims that, beyond the purpose of selling the product and brands themselves, a whole range of meanings are added when companies present themselves to the global market:

> Certainly food is nourishment and cars are for transportation, but above this, products and brands are a communication system for values, social distinctions, and power. Products are extensions of the self and are important in establishing a personal or household identity and in self-presentation.

The implications of this claim incorporate his view of the pervasiveness of culture and its relevance to the individual as well as to advertising:

> Culture is not restricted to objects in museums. Culture pervades our daily lives through the products we use, the buildings we live in, and the television programs we watch. [...] The dominant culture of a country relates to basic cultural assumptions, history, national, ethnic, religious and social values. The dominant culture pervades the other levels of culture, such as organisational cultures (corporate identity), consumer (sub)cultures, and the material culture of products, design, architecture, and advertising. (Van Raaij 1997: 261)

Further to this, as Figure 1 shows, the USA represented in the symbolic icon of the Statue of Liberty, toasts with a glass of Italian red wine (see also the verbal text – Vino 2009 [Wine 2009]) one of the traditional images of Italy, the Leaning Tower of Pisa located in presumably Italian countryside. The message of appreciation of the wine is implicit.

Figure 1. Italian wine fair USA.

Figure 2. Chianti in the USA.

As Heslop *et al.* (1993) claim, a strong country-of-origin image of a product/brand makes it acceptable for the brand to be advertised with its original background and less towards the target group country. However, in Figure 1, this kind of acceptance is created through the combination of the two culturally iconic images where the 'host' country is seen to be offering a toast to the other. The Chianti advert (Figure 2) further offers a tongue-in-cheek message ("America has discovered an Italian") with the intention of paying homage to Italian excellence in terms of wine and relying on knowledge of the world regarding the discovery of America with red flags (wine) representing ship sails coming out of bottles masts.

3. National identity and communication: an overview

As far as product advertisements are concerned, a country can be identified in a number of more or less subtle ways, relating to its geography or its national colours, thus foregrounding the country of origin rather than the product itself. Figure 3 identifies Italy in the map of the country, purposefully coloured yellow to represent the colour of the Parmesan cheese advertised, as well as the icon representing the product itself, positioned exactly where it is manufactured in the Parma area.

Figure 3. Parmesan cheese advert.

Figure 4a. Fiat 500.

Figure 4b. Italian wine.

In the Fiat ad (Fig 4a), on the other hand, place is represented by the Italian flag in the colour trail along the road, together with the green, white and red coloured Fiat cars, while in Figure 4b, the colours of the Italian flag are likewise represented in the green, white and red contents of the wine glasses positioned throughout the map of Italy, with nothing to say, however, about the product itself. The message is reinforced in the verbal "Salute!" [Cheers!], a typically Italian way of toasting people when drinking wine together. Indeed, Kelly-Holmes (2005: 36) notes how

> [T]he presence of a word or phrase from the language associated with the country of origin reinforces both the visual and/or aural texture of the advertisement, working with the paralanguage, the graphics, the particular scenario, use of colour and so on, to reinforce the message.

As far as luxury items are concerned, and especially perfume, the country may be represented in the brand name of the product: Roma (Biagiotti); Acqua di Parma; Via Condotti (Lancetti); Piazza di Spagna (Roccobarocco); Tuscan Soul (Ferragamo), while it is worth noting that Estée Lauder, an American company, names one of its products Tuscany, giving an example of a kind of linguistic fetish which associates the perfume product to Italy.

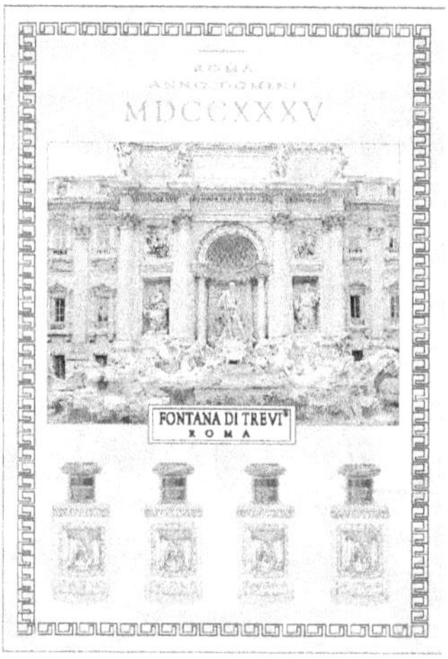

Figure 5. Fontana di Trevi perfume.

We can also note how there is a clear intermingling of brand, product and place where, for the Fontana di Trevi perfume (Figure 5), not only an image is presented of the actual place, but also the name of the city, Roma, reinforces the message, a feature common to other luxury and fashion products (see also Prada, Milano). Thus clearly a foreignising strategy is adopted for the foreign market. On the other hand, Sicily (Figure 6) is used as a brand name due to its soft /s/ sound rather than a harsher one in the Italian pronunciation of 'Sicilia'.[2]

2 Personal communication from Donatella Carrirolo, Euroitalia, at the time distributor of Dolce e Gabbana perfume.

Figure 6. Dolce & Gabbana Sicily perfume.

Furthermore, we can begin to move towards other implicit meanings conveyed through the images and message (the Italian actress, Monica Bellucci), colour (black and white photo apart from the yellow liquid contents of the perfume bottle, the brand and product name as well as the slogan) and shortest of verbal texts ("Feel the passion"). This print ad is a frame from the TV commercial directed by the Italian film director Tornatore, which has a Sicilian setting, depicting a young widow at the death of her husband and her attraction to another Sicilian and his to her. The emphasis of the TV ad is still on the concept of place. It utterly exudes cultural images and references to Sicily (the musical soundtrack, the young man wearing the traditional Sicilian *coppola* cap, the portrayal of mourning at a Sicilian funeral). The slogan under the packshot ("Feel the passion") reinforces the emotion intended.

Figure 7. Lavazza Italian Espresso.

Passion in the previous example is a stereotypical Italian trait often exploited in advertisements to emphasize the intensity of Italians' feelings in whatever they do (whether they make pasta, drive cars, or drink coffee). Figure 7 combines and highlights this emotion (here passion all for a cup of coffee!!) with the sense of place, and particularly an ancient Italian building creating romantic surroundings, together with the colour red in the woman's long flowing dress and a red background for the packshot and bodycopy in the vertical strip to the right of the image. Indeed, the introduction of a romantic element – handsome dark-haired men and beautiful dark-haired women (not forgetting the Italian epitome of beauty in Monica Bellucci's testimonial in the perfume ad) is a common feature in cosmetic ads.[3] Likewise, some adverts of Parmesan cheese (e.g. Figure 8) exploit the value of elegance and style that Italy is famous for, with a wedge of cheese imitating a stylish handbag.

3 On the other hand, Barilla ads in the USA portray typically blonde American women.

Figure 8. The Italian food collection.

The above examples confirm that national/local culture and identity are relied upon when it comes to the representation of products and nations in the world of global advertising. The following sections will therefore investigate the main strategies which are commonly deployed to enhance a company's competitive advantage.

4. The ceramic industry

Nowadays, when thinking of Italian products, not only wine, cheese and perfume come to mind, but also ceramic tiles. Typically, international buyers tend to identify the Italian ceramic tile industry with some specific geographical areas: one of them being the Sassuolo-Scandiano area, where most of the Italian ceramic tile companies are located. In this case reference to place is used to enhance the value of the brand knowledge and stress the added value provided by the presence of a network of businesses (business cluster), which generates excellence and a higher capability to develop new products. Within this perspective, firms can develop place of origin-based branding strategies, such as using product names that recall place or other local-based elements, or the CI label in addition to the brand name.[4]

A company's location is not viewed as a mere geographical parameter, but as the result of specific social, cultural and institutional elements that support the economic action of interdependent networks specializing in the same or related sector, and give rise to specific business clusters conveying the positive image of a system of firms and people with the right expertise for designing, making and selling a product. The Italian ceramic industry, one of Italy's key manufacturing sectors, is composed of companies specializing in ceramic tiles, ceramic sanitary ware, ceramic tableware and refractory materials. In the 1970s, when the ceramic tile industry in Italy was in the earliest stages of its development, it generated mainly small businesses producing at craftsman level and supplying only local markets. Then, a key factor was introduced in the sector and sharply accelerated the production cycle: fast single-firing technology. This prompted a dramatic increase in production and a huge improvement in productivity which rapidly led to the saturation of the domestic market. Therefore, the firms began to tackle the problem of excess capacity by trying to enter the closest foreign markets (Germany and France). However, the penetration of these markets was anything but easy: barriers were very high, due to customers' established preferences for local products and their almost complete lack of awareness of Italian companies' brands. An effective solution to this problem was given by the creation, in the 1970s, of the 'Ceramic Tiles of Italy' (CTI) brand name for the Italian ceramic industry world-wide. The CTI trademark was protected by international copyright and could be used uniquely by Italian companies belonging to Assopiastrelle (Confindustria Ceramica), the association of Italian ceramic tile producers, founded in 1964. The label was meant to evoke a sense of tradition, quality, innovation and creativity. It safeguarded the reputation of Confindustria Ceramica member companies and their products and embodied the desire of Italian ceramic tile producers to promote an awareness of Italian ceramic tiles amongst international users. However, since the CTI brand was used

4 Cf. for instance, the CTI or CI labels which will be dealt with in one of the ensuing paragraphs.

solely for ceramic tiles, a new brand was later developed – Ceramics of Italy (CI) – which identifies ceramic tile, sanitary ware and tableware produced in Italian factories.

Figure 9. The logo of *Ceramics of Italy*.

5. Objectives, materials and methods

We will now examine a corpus of Italian company websites and one of multimodal ads in order to discover how the Italian-ness of both the company and its product(s) is conveyed verbally and non-verbally to a global audience through the medium of English, albeit with foreignizing and/or domesticating strategies in the choice of lexis to emphasize the identity of the manufacturer and product. The analysis covers a range of typical Italian products, with a view to identifying those features that contribute to paving the way for each company's self-representation to a foreign audience. Accordingly, attention will be focussed particularly on advertising strategies and on the brand and product names.

Moreover, as far as the products (and websites too) are concerned, the aim of the analysis is to show how for some Italian products, advertising is moving towards a standardization in setting (place) through verbal, visual and acoustic attributes, albeit with a localized orientation to maintain country of origin. This brings us to the question of formal or functional equivalence in global advertising where, by formal equivalence we mean a complete standardization with identical campaigns in different countries (e.g. Coca-Cola campaigns in

American English), and by functional equivalence, campaigns that may have an identical mission or communicative objective, but employing different concepts and executions to convey the 'same message'.

As far as food, wine and luxury products are concerned, it is possible to rely on a wide range of sources, given the multi-faceted nature of the sectors involved. In fact, the adverts of the Italian products presented in the analysis below were collected from corporate websites, web searches, video launches of products at fairs, TV ads, paper press, following as a simple criterion for selection the fact that the products or brand names are typically associated with Italy: pasta (Barilla, Giovanni Rana); coffee (Illy, Lavazza); perfume (Dolce & Gabbana), automobiles (Fiat, Alfa Romeo, Ferrari).

By contrast, regarding the ceramic tile industry, reference was made mainly to corporate websites. In particular, our attention focussed on Lea Ceramiche, one of the brands of Panariagroup Industrie Ceramiche S.p.A, and on the Marazzi Group. Lea Ceramiche is characterized by a range of upmarket products. Lea Ceramiche and the Marazzi Group can be considered as two of the most dynamic and productive brands of the Modena ceramic district. These companies design and produce floor and wall ceramics for all kinds of settings and have a far-reaching national and international coverage, with a capillary presence in more than eighty countries in the five continents around the world. The analysis centred on the companies' website, one of the principal and most immediate tools used by companies to generate awareness about their products, services, events, news and so on. Unlike other text types, corporate websites,

> in which the author uses evaluative language to position himself within the norms of a discourse community or an institutional image, [...] create a profile of the company which distinguishes it from its competitors. (Salvi/Turnbull 2007: 244)

Accessing the English version of the websites, a potential customer is confronted with the navigation bar which, in turn, leads to the different products. In particular, for the purpose of the present analysis, attention has been devoted to those parts of the two websites which deal with the different collections offered by the two companies.

As far as the method of analysis is concerned, in the first place reference was made to the work of de Mooij (2005), which highlights the focal points of an analysis of advertising strategies (cf. Figure 10). A brand name can be associated with a number of elements, including relevant features that reflect self representation, such as place (Italy), product (our corpora), visual image (the product and cultural reference to Italy) and occasions, moments and moods (the setting of the advert).

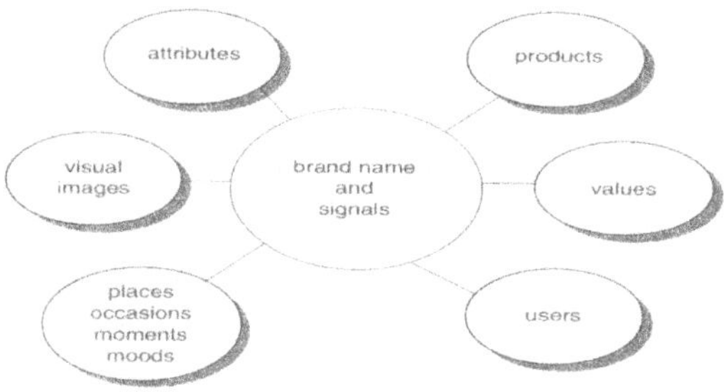

Figure 10. de Mooij – Elements of brand association network (2005).

Other levels of analysis (Chiaro 2004, 2007, 2009) were useful in focussing on the medium of communication: the verbal – including implicit and explicit reference to cultural identity and values; the visual – including colour, and acoustic – including music, songs, foreign accent. More specifically, Chiaro highlights the attributes of Italianness in agro-food advertisements and classifies them in turn (verbal, visual and acoustic) according to such themes as family and tradition, authenticity and naturalness, stylishness, good life, art and beauty and, finally, passion (2009: 40-41), all of which, as will be seen in the following sections, are features present in our corpus. In the corporate website, which could be compared to a shop window for the organisation, where all kinds of available information about the firm are displayed, the analysis focused in particular on the names chosen for the different collections and products, since the language employed does not limit itself to mirroring objective realities, but guides the stakeholders to the appreciation of the company's credibility, stressing meanwhile its professional competence and expertise (Salvi /Turnbull 2007: 243).

6. The analysis

As Kelly Holmes (2005:49) claims, "linguistic fetish is not always confined to lexical items. Syntax, style, accent, dialects – all of these are exploited in advertising and can stand alone as an independent source of meaning, and in the process give meaning to a particular product in terms of cultural competence."

Giovanni Rana pasta UK uses a TV commercial with a romantic situation between a young English man and an attractive Italian neighbour, Sofia, who, after being enticed into his flat under the false pretences of a burnt dinner for a guest, cooks him some Giovanni Rana pasta, describing it in her exquisitely Italian accent. The commercial ends with the sound of Italian opera music blasting the scene as they sit down to eat (she is the intended guest!).

Products may travel simply exploiting the acoustic attribute, again with the distinctly Italian accent of the wine producer, Zonin (Figure 11). His company uses a short video clip to portray groups of people of different ages drinking sparkling wine in an unidentifiable international setting (e.g. a modern city in Europe or the USA), but the country of origin of the wine is recognizable from the voice of the Italian wine producer himself when he pronounces the only words in the commercial, "Enjoy the simple moments of life", as used also in the print version in Figure 11. In her studies on the agro-food sector, Chiaro (2007) distinguishes between two kinds of imagery of Italy – traditional and modern, where the wine advert below (Figure 11) tends to the latter in a form of standardization, while others project a more traditional image (Italian vineyards in front of magnificent villas or typical coastal areas or villages of sun-baked Italy).

Figure 11. Zonin Italian wine.

What is particularly interesting to note about the Zonin ad is the promotional strategy of the company that emerges with respect to its website. While the latter foregrounds the company and adopts a more indirect and formal stance with its readers (referring to the family in the 3rd person) and also focussing on family tradition, historical Mediterranean tradition and locality (Turnbull 2007: 26-7), the printed and TV ads depict a member of the family addressing directly potential customers with another kind of message that relates to and implies modern lifestyles.

Global pasta ads for Barilla adopt a very sender-oriented approach, invariably set round a kitchen table focussing on the concept of the family. It is worth noting how the Barilla TV commercial for the American market not only reinforces the message of family values with the transposition of the same setting, but also places emphasis on the country of origin, through the soundtrack – Andrea Bocelli singing in the background a familiar Italian song, and an Italian guest at the table speaking English with a recognisable Italian accent. Some settings of Barilla USA ads also recall the Tuscan countryside with a member of the family returning home for a meal. The marketing strategy here is to reinforce the country-of-origin effect when the product is actually manufactured in the USA. Furthermore, the Barilla concept of family values are clearly identifiable in a commercial which takes place in Italy with an American wife preparing lunch for her sister (both are blonde) who has just arrived from the airport on a visit. A dark handsome Italian neighbour joins them and their children at the dinner table. The final message is: "Barilla, the choice of Italy". Thus we can identify the following attributes: acoustics (Andrea Bocelli singing); verbal (the children call out *zia* [auntie] when their aunt arrives and the dark handsome Italian comments on the pasta dish being *perfetto* [perfect]; visual (the Italian landscape, the dark handsome Italian male, an obvious attraction between him and the fair-haired aunt; and in conclusion, self-glorification with the final "the choice of Italy".

Now moving to another product, Fiat 500 US TV commercials (2011-12) appear to be less sender-oriented but addressed to young people, conveying other messages, such as the functionality and performance of the car. The Fiat campaign includes testimonials of celebrities, such as Jennifer Lopez driving a new Fiat 500 around New York, not only in the modern streets, but around the poor area she was brought up in. Thus we can note that the values conveyed are applicable further afield, as expressed in the verbal message we hear Jennifer Lopez give (her singing also accompanies the clip): "Here, this is my world. This place inspires me, to be together, to stay sharper, to think faster. They may be just streets to you, but to me they're a playground." While there is more emphasis on the functionality and performance of the product from the visual, the caption on the screen at the end "Life is best when driven" is a combination of not only the product but also a wider range of modern values relating to life presented in a message directed at young people.

The Fiat 500 USA campaign used a number of other marketing techniques to promote the product Fiat. In the advert below (Figure 12), there is a predominance of red (the car and the internal part of the Fiat logo), indicating the passion in Italian blood that we saw earlier on in the Illy coffee advert (Figure 5), re-inforced by the verbal message: "And begin a torrid affair".

Figure 12. Fiat 500.

This Fiat ad, and others in their campaign, fits into De Mooij's value structure map (Figure 13). Values are pleasure, independence (Jennifer Lopez feeling free), achievement, performance. Benefits are personal enjoyment, imaginative, daring (in some Fiat TV commercials a man and woman challenge each other to race madly round a racetrack), beauty (in another, a young man stands dreamily at the side of the road, stares at a sexually desirable girl, but then she transforms into a black Fiat 500), and finally attributes, strong motor, quality, design, and so on.

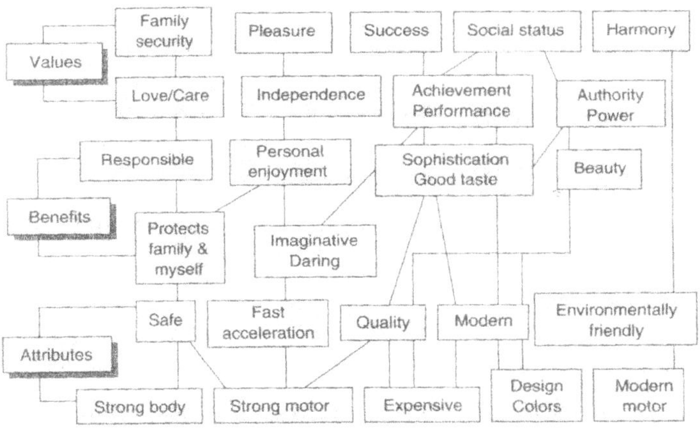

Figure 13. De Mooij's value structure model.

6.1. Lea Ceramiche's and Marazzi Group's portfolio of products

As noted earlier, corporate websites could be compared to a shop window for organisation, displaying all kinds of available information about the firm and its products. Focusing on the range of products offered by the two companies under scrutiny here, in the first place attention was devoted to the names of the products, where both foreignizing and domesticating strategies were applied in a mixture of noun phrases, either retaining the Italian language, or incorporating both English and Italian words, as the examples below show. Some product names were probably chosen on the grounds of the name of the material that the tile collection tries to 'imitate'. For example, the tile collections which try to imitate the wood effect are called *Bioessenze* (Lea Ceramiche), where the prefix *bio-* conveys an idea of something belonging to the semantic field of nature and life, as is the case of wood. Similarly, another collection of ceramic-wood by Marazzi is called Treverk, which evokes some of the consonant and vowel sounds of a particularly hard kind of wood ('teak' or 'tectona').

The collections which imitate the stone effect are called *Basaltina Stone Project* and *Arenaria* (Lea Ceramiche), the former combining *basalt* and the English word *stone,* the latter deriving from the scientific name of a stone. *Pietra di Borgogna* (by Marazzi) reminds us of the simplicity of materials from the past, while *Pietra di Noto* evokes the Sicilian baroque style and is characterized by sophisticated elegance and timeless beauty. Similarly, *Graniti*, *Porfidi*, *Le Pietre* and *Cotto Antico* (Marazzi Group) directly remind us of the peculiar features of the types of stones they set out to imitate.

In addition, *Digiquartz* and *Tecnoquartz* (Lea Ceramiche), imitate the quartz effect, in a similar way to *Multiquarz* by Marazzi, which offers porcelain slabs with a tactile look and strong shade variation, with surfaces perfect for outdoor use. In these cases, as with the *Basaltina Stone Project,* we witness instances of code-mixing, which combine Italian words with English ones.

The product name *Slimtech* (Lea Ceramiche) is made up of the two English words *slim* and *technology* (which is often abbreviated as *tech,* as happens for example in *hi-tech*). In this case, the product name is a sort of description of the tile itself: in fact the Slimtech collection is characterized by ultra-thin porcelain slabs. The same applies to the *Silkstone collection*, inspired by stone and silk. Once again, the use of words of English origin highlights the global dimension of the company.

Another product by Lea Ceramiche is called *Re-evolution:* this product name may conceal a dual reference. Firstly, it conveys the idea of change, of evolution, or better, of a new evolution, which, we may infer, is more innovative and modern with respect to the previous one. Secondly, it recalls the idea of revolution too, so of something which will modify the existing *status quo.* With this product name the company may well want to underline that it is committed to innovation, modernity and research. The same implications also apply to the *EvolutionStone* by Marazzi, which is the original interpretation of seven valuable natural stones: Serena, Piasentina, Ardesia, Malaga, Luserna, Quarzite and Pierre Bleue.

The names of three wall covering tile collections by Lea Ceramiche are: *Très Jolie, Audrey* and *Kensington.* The first name conveys the idea of something pretty, nice, lovely; moreover, the lexical choice of the French gives to the product an aura of sophistication. The same happens for *Audrey,* which is a clear reference to the well-known actress Audrey Hepburn and for Kensington, which recalls the atmosphere of English residences. These three product names are synonyms of sophistication, style and design. The company wants to emphasize that its products have a high design content and may contribute to giving life to a stylish location, if used as wall coverings. We should also underline the fact that the product names reflect the product characteristics and styles too, in their colours and decorations. Moreover, it is interesting to note how the company does not confine itself to words or terms denoting concepts or values pertaining just to Italy, but to a wider area indicating its tendency to present its products as being relevant to more than one culture. This is attested by the names of three other collections of tiles by the Marazzi Group, *Paris*, which conveys the appeal of a timeless city, with a rich and luxurious style; *Soho*, with a definite, minimalist and fashionable style; and *Brooklyn* ceramic tiles, inspired by concrete.

The names of the above mentioned collections clearly display the adoption of domesticating techniques; the names of other products, on the other hand, seem to evoke references to the Italian and classical cultural heritage, and, as such, a foreignizing attitude: *Agorà,* which reinterprets the purity and essentiality of marble in a modern style. *Caracalla*, which expresses the purity and essential nature of marble. The *Policromi* series is characterized by veins and shaded surfaces with a rustic look that recall six different types of marble. Finally, the *Dolcevita* series, which focuses on terms of elegance and sophistication and is clearly associated with Italy's renown for style; the *Etruria* collection which evokes the elegant appeal of traditional terracotta and clay tiles.

7. Discussion of findings

As shown by our analysis, self representation emerges through the display of products and the publication of company websites on a global scale. This relies heavily, on the one hand, on the precise reference to and identification of people and places and, on the other, on *global* advertising strategies that highlighted culture and country of origin. Reinforcement and/or redundancy of the self-representation message is clearly identifiable through the various channels of communication (verbal, visual, acoustic attributes). The words and pictures, in the case of our own corpora, maintain elements of foreignization that immediately associate the product or the company with its country of origin.

In discussing Venuti's foreignizing and domesticating strategies (1994) with respect to the foreign words included in a British advertisement for Dolmio pasta sauce, Kelly-Holmes (2005: 17) coins the term 'domesticated foreignness' to emphasize the fact that it is not simply an Italian product that is being advertised, but a British idea of it. She justifies her argument by quoting Venuti's claim (1994: 220) in relation to the translation of advertising:

> when [the] products are foreign, the significance must be domestic but its reverberation will be intercultural: a translated ad can simultaneously create or

> revise a stereotype on a foreign culture, while appealing to a specific domestic constituency, a specific segment of the domestic market.

As far as the ceramic industry is concerned, it is possible to see that in addition to the country-of-origin effect, conveyed by the 'Made in Italy' and CTI or CI brand labels, the names of the products reveal instances of creative word formation, achieved by means of affixation (-*quartz*, *bio-*, etc.) and word compounding (*Basaltina Stone Project*; *Slimstone*; etc.). In addition we can find Italian words charged with a positive connotation, both in terms of references to Italy's cultural heritage (the *Policromi*, *Pietra di Noto*, *Caracalla*, *Etruria,* etc.) and elegance and sophistication (*Dolce Vita*). Finally, the references to foreign places and names also attest to the adoption of domesticating strategies, as a means of reaching out to international stakeholders, which may be attributed to the desire of the company to underline its international dimension and to attract clients on a global scale.

Some TV and print ads may well present a 'more internationalized' person and/or situation (people drinking wine at a party, a family or friends sitting round a breakfast or dinner table), yet one or more of the verbal, visual and acoustic components will inevitably betray the country of origin. On the other hand, in a more traditional approach (e.g. the image of Italian vineyards, Italian architecture), the country-of-origin effect is reinforced in the image itself.

Van Raaij (1997: 266) claims that while marketing strategies are identical all over the world, global communication is sender-oriented, which, indeed, our findings regarding self-representation clearly confirm:

> In a global strategy, the communication is rather sender-oriented. The mission, proposition, concept and execution are identical over cultures. Note that a global strategy does not mean here 'Think global, act local!', i.e., that different communication strategies are used locally, integrated in a global strategy. In adaptation and differentiation strategies non-identical concepts and executions may be used, although with the same mission and proposition. In a local strategy, a completely different (unique) approach is taken for each culture.

Thus, in our discussion, it is important to distinguish between common purpose (the selling of the product) and the strategies peculiar to relaying the added value of self-representation of the company in a global context without losing trace of the country of origin.

8. Conclusions

There are numerous products for which Italy is renowned and enjoys worldwide fame and the following are just a small example: leather goods, textiles, fashion, sunglasses, ceramics, glass, and last, but not least, its cuisine and its wines. The evidence provided by this small-scale study suggests that in the two different areas under scrutiny, identity is established in relation to a sense of place, by referring to local (Italian or foreign) places, thanks to the deployment of a variety of foreignizing or domesticating strategies. In particular, the analysis of the websites of two well-known tile manufacturers has shown that when it comes to the names of products and collections, alongside Italian words and references to the Italian culture and traditions, foreign words and cultures are also mentioned, as if to imply that the sense of Italianness is characterized both by strong ties with the country's set of traditions and by an innovative, multi-cultural drive. This in line with the findings of previous studies (Salvi 2012: 112), according to which the language of corporations is transformed and modified in order to suit the language norms of local communities, with the adopted lexicon bearing traces of phenomena of contamination, as a result of globalization.

Our analysis has shown how both the language of advertising and the promotional language identifiable in company websites have a common aim in creating an implicit and/or explicit country-of-origin effect. This is achieved by exploiting a combination of verbal, visual and acoustic attributes according to the medium in which the message is conveyed. The co-presence of people culturally related, so to speak, to the product and who are in some way participants of the promotional message (often with reference to family tradition, to their commitment to the manufacturing and quality of the product) is central to the underlying communication strategy. Indeed, de Mooij (2005: 4-5) reinforces this approach by pointing out that global products are advertised through local people and local images: "Markets are people, not products. There may be global products, but there are no global people. There may be global brands, but there are no global motivations for buying those brands." Moreover she describes many global advertisers as being product rather than market oriented, in which case the country-of-origin effect, or the reaching out to international customers and their different places of origin is foregrounded in some way.

And finally, our analysis also aimed both to confirm and disconfirm Myers's (1999: 61) claim that: "The brand image can be global, but the words and pictures will usually change from market to market." In this light, in a localized world, there may be either a redundancy or omission of cultural features, that need no self-representation on the part of the company. On the other hand, in the vision of a glocalized community, implicit or explicit reference to cultural identity is essential in associating the product to its country-of-origin manufacturer, and at the same time include in the accompanying descriptive language (name of the product) concepts and values relevant on a worldwide scale.

Selling who we are and what our product is, linking product and people (be they manufacturers or consumers) thus appear to be inseparable in the advent of globalization.

References

Chiaro, Delia 2004. Translational and Marketing Communication. A Comparison of Print and Web Advertising of Italian Agro-Food Products. *The Translator* 10/2, 313-328.

Chiaro, Delia 2007. A Question of Taste: Translating the Flavour of Italy. In Garzone, Giuliana / Ilie, Cornelia (eds) *The Use of English in Institutional and Business Settings An Intercultural Perspective.* Bern: Peter Lang, 57 - 79.

Chiaro, Delia 2009. Italians Doing it on the Web: Translating and Selling Italianness, *VIAL* 6, 33-49.

de Mooij, Marieke 2005. *Global Marketing and Advertising. Understanding Cultural Paradoxes*. London: Sage.

Hare Hansen, Jens 1997. Translation of Technical Brochures. In Trosborg, Anna (ed.) *Text Typology and Translation.* Amsterdam: Benjamins.

Heslop, Louise A. / Papadopoulos, Nicolas 1993. But Who Knows Where or When? Reflections on the Images of Countries and Their Products. In Papadopoulos, Nicolas / Heslop, Louise A. (eds) *Product-Country Images: Impact and Role in International Marketing*. New York: International Business Press, 39-75.

Jaffe, Eugene D. / Nebenzahl, Israel D. 2001. *National Image and Competitive Advantage: The Theory and Practice of Country-of-Origin Effect.* Copenhagen: Copenhagen Business School Press.

Kelly-Holmes, Helen 2005. *Advertising as Multilingual Communication*. Basingstoke: Palgrave.

Melewar, Tengku C. / Karaosmanoglu, Elif 2006. Seven Dimensions of Corporate Identity. A categorisation from the Practitioners'. *European Journal of Marketing*. 40/7: 846-869.

Myers, Greg 1999. *Ad worlds: brands, media, audiences.* London: Arnold.

Salvi, Rita 2012. Textual Equivalence in the Translation of Specialized Languages: Discourse Adherences and Dissimilarities. *Altre Modernità* 11, 102-115. <http://riviste.unimi.it/index.php/AMonline/article/view/2465>.

Salvi, Rita / Turnbull, Judith 2007. Corporate Websites: Voices of Authority. *Textus* XX/1, 243-264.

Turnbull, Judith 2007. Language and Culture in Corporate Websites. In Salvi, Rita / Bamford, Julia (eds) *Business Discourse: Language at Work.* Roma: Aracne Editrice, 17-34.

Van Raaij, Fred W. 1997. Globalisation of Marketing Communication? *Journal of Economic Psychology* 18, 259-270.
Venuti, Lawrence 1994. *The Translator's Invisibility.* London and New York: Routledge.

JUDITH TURNBULL

Building, Enhancing and Defending Reputation in a Corporate Website

1. Introduction

A corporate website is an open invitation to the world to enter a virtual space, unbounded physically and geographically, where a company can, in absolute freedom, creatively construct and project its characteristic and distinctive identity. With the evolution of the concept of Corporate Social Responsibility corporate identity has become a complex phenomenon as modern business has increasingly embraced economic, legal, ethical and philanthropic responsibilities. However, the term 'corporate identity' is used sometimes in a rather vague manner and confused with other terms as if they were synonyms (Abdullah 2009, Löwensberg 2006) and therefore they require some clarification.

Corporate identity is the self-presentation of an organisation, how it views itself and how it wants to be seen. This means not only the visual identity of a company, through its logo, colour and any graphical element that expresses the essence of a company, but also its business strategy, organizational culture and structure and especially its "espoused values and philosophy" (Balmer *et al.* 2007: 8).

The reverse side of identity, so to speak, is the concept of image, which can be considered a reflection of an organisation in the eyes or minds of its publics. It is how an organization is perceived by others, its stakeholders or constituencies, as they are referred to in management. This means that identity is:

> not always neatly 'located' within the company. In other words, corporate identity – rather like individual identity – cannot be said to be solely based in some core essential entity that simply emits its distinctive character. Instead

> identity is formed as much in the mirror of another's perceptions, with this reflection being as much an influencing factor in how a corporation behaves and asserts its identity. (Balmer *et al.* 2007: 10)

There is also a third dimension to identity, namely reputation, which is the sum of a series of images over a period of time which together form an overall, affective impression. Since we often talk in terms of a good or bad reputation, it also involves social evaluation. Whilst the image of an organization can be compared to a snapshot of the company at a particular moment in time, reputation is necessarily built up over time as a result of consistent performance. As Henry Ford said, "You can't build a reputation on what you are going to do". Reputation is vital for stakeholder trust and is considered 'a key source of distinctiveness' and obviously both these factors will gain the corporation competitive advantage, though reputation requires constant and harmonious measures in its management (Abdullah 2009: 172).

Generally identity in virtual space is considered more unstable, creative and volatile than real life identities. However, Benwell and Stokoe conclude that virtual communities recreate conditions of real life rather than forge radically new ways of conceiving relations and identities (2006: 278). This is particularly true of a corporate website, as companies need to remain firmly rooted in the real physical world. In real life, time and space "provide the network of relations by means of which we can situate objects, events or states of affairs" (Galton 2011: 695) and thus make sense of the world about us. Bearing in mind the channel of communication and the status of a multinational operating all over the world with its concomitant reputation, time and space act as fundamental coordinates in the construction of corporate identity and reputation.

This chapter will therefore examine how a multinational uses these coordinates to build, enhance and, if necessary, defend its reputation in one of the most important channels of business communication available today, its corporate website. It will not deal with crisis communication when a company's reputation has been tarnished and desperate attempts are made to salvage it. Instead it will investigate how a multinational 'manages' its reputation on a day-to-day basis through spatial and temporal frameworks and references in a particularly sensitive area, sustainability, which can easily put a multinational's reputation at stake.

The study has been carried out on the website of Unilever, an Anglo-Dutch corporation, which produces some of the world's leading consumer products in food and beverages, personal and home hygiene. Unlike other multinationals, Unilever has not been the centre of particular scandals or controversies, though big corporations generally are frequently accused of being ruthlessly market-minded, of aiming purely at maximising profits and encouraging irresponsible consumerism. In November 2010 Unilever published its Sustainable Living Plan in which it outlined how it proposes to work towards sustainable growth. It was accompanied by a section of Frequently Asked Questions which tackled many of the controversial aspects of multinational business[1]. These FAQs are the subject of this study as they provide an opportunity to see how Unilever positions itself in relation to its publics in the face of such challenging issues.

2. Identity and reputation in corporate communication

Corporate communication consists in all the messages from both official and informal sources through a variety of channels by which a company communicates its identity. And nowadays a corporate website is one, if not the most important, of these. It acts as a virtual meeting point for people all over the globe and offers a rhetorical space in which the company can share and discuss its ideas in a public sphere.

It is often said that corporate communication needs to be "harmonized as effectively and efficiently as possible" (van Riel 1995: 26) in order to present a single homogeneous picture of the company in both internal and external communication. This is achieved through the use of leitmotivs, characteristic and coherent themes and style that

1 The Sustainability Living Plan is accessible on line at <www.unilever.com>. However, the FAQs accompanying the document, which were downloaded on February 5, 2011 for this study, are no longer available.

construct and form the corporate identity (Löwensberg 2006: 239). Companies and products take on a 'brand personality' founded on a set of core values (Koller 2007: 114) with which their stakeholders can identify. However, working in an international context like a website raises the question of how values and social practices, which are inherently culture-specific, can be negotiated in a virtual world which transcends national boundaries (Turnbull 2011).

The importance of corporate personality has been stressed as a key reputation construct to measure both the image and the identity of organizations from the perspective of their behaviour (Abdullah 2009: 172). However, whereas a good communication programme can build a strong identity, it is not sufficient to build a strong reputation, though it can help to reinforce and promote it (Gray/Balmer 1998: 696).

Indeed, as many stakeholders are becoming more and more interested and concerned about the way corporations go about their business, corporate reputations are subject to attack. In particular, the emergence, on the one hand, of the ethical consumer and his/her growing power to influence corporate behaviour and, on the other, the ethical investor who may fear the consequences of a damaged reputation, means a corporation must manage relationships with its constituencies carefully (Pruzan 2001: 51-53).

Pruzan suggests that there are two complementary perspectives on the concept of corporate reputation and its relationship to success and credibility. Firstly, there is the pragmatic approach to reputation which aims to build up the corporate image. It makes sure the company cannot be accused of breaking the law or behaving unethically and it is sensitive to the demands of customers, shareholders, local community and so on, focussing on better information and the manner in which the corporation communicates externally.

Alternatively, there is what he calls the reflective perspective which is more concerned with the inherent 'character' of the organization rather than its image and outward appearance. Corporations ask themselves who they actually are, what values they support and what their obligations are as an organization, looking inwards at its very nature, integrity and identity. The answers to these questions form the 'guiding philosophy' of the organization.

The distinction between the two approaches has been reiterated recently, though in more pragmatic terms, as instrumental and ethical (Shinkle/Spencer 2012). However, it is very difficult to distinguish between the two approaches, as can be seen in the following statement which appears in the sustainability section of the global version of the website and shows how Unilever has apparently adopted Pruzan's reflective approach.

(1) **Values & standards**
We aim to be a trusted corporate citizen wherever we operate in the world, respected for the values and standards by which we behave.
Our shared values
We have a set of shared values. Our values guide the way in which we do business and influence the way we think and act. It is by putting these shared values into everyday working practice that we can operate successfully as a company.

It stresses how the corporation has integrated its values and way of thinking into its activities and working practices. The choice of the word *citizen* is particularly significant because citizenship carries with it both rights and responsibilities to the community; a citizen should work towards the betterment of his community and, in the case of Unilever, this community is *wherever we operate in the world.*

Ethical identity is seen as being formed relationally between parties through business and social exchanges (Balmer *et al.* 2007: 8). Indeed, it is a well-established idea in the literature, that "identity, rather than reflected in discourse is actually constituted actively, ongoingly and dynamically in discourse" (Benwell/Stokoe 2006: 4). Coupland and Brown (2004) also suggest that organizational identities are authored by a 'dialogue' between the company as 'insiders' and stakeholders and the general public as 'outsiders'. Nowadays this dialogue can take place in a blog or forum for a direct exchange of ideas, though the Unilever website has neither one nor the other.[2] Communication is purely unidirectional, from the company to the visitor. This is probably because blogs and forums require a lot of maintenance and monitoring – an unfavourable blog or comment can destroy a reputa-

2 However, the website now has links to Facebook and Twitter.

tion in a very short time. In contrast, the discourse and identity work in 'Frequently Asked Questions' can be controlled and "more rigorously policed and carefully delineated" (Benwell / Stokoe 2006: 278).

FAQs are a typical genre in websites, a special kind of virtual space which tries to simulate a discussion between insiders and outsiders. Generally FAQs are used as cues to give further information on a topic or to address an issue from a different point of view. They may also be seen to reflect the perception of the firm's identity, that is, the corporate image in the public mind, as they are *apparently* the questions that the public actually asks the firm. However, on such a fundamental topic as sustainability the company also anticipates and responds to many of the criticisms which may be levelled against it on issues concerning social, economic and environmental issues. By responding to questions that could potentially threaten its reputation, it acknowledges the concerns of its stakeholders and tries to forestall attacks and at the same time enhance its reputation. In this way we can view FAQs as acts of reputation management.

3. Corpus and methodology

The corpus is made up of the Frequently Asked Questions accompanying a document on the company's sustainability plan, the Unilever Sustainability Living Plan. It is made up of ten subject areas, touching upon both the environmental and social aspects of sustainability: The Plan, Hygiene, Nutrition, Greenhouse Gases, Water, Waste, Sustainable Sourcing, Better Livelihoods, People, Paper and Board, covering 111 different topics, some with multiple questions, for a total of 127 questions and 15,353 words.

Subject area	topics	words	Subject area	topics	words
The plan	11	2600	Waste	10	1236
Hygiene	12	2949	Sustainable sourcing	13	1586
Nutrition	10	1010	Better livelihoods	13	1465
Greenhouse gases	12	1142	People	4	605
Water	11	953	Paper and board	15	1807
TOTAL				111	15353

Table 1. Composition of corpus.

In view of the small size of the corpus a discourse analysis approach has been adopted which tends towards a Critical Discourse Analysis approach given the size of the business of Unilever and the power relations inherent in the dominating position of a multinational. ConcApp and WMatrix software were also used to draw up quantitative data to support the analysis.

The FAQs have also been analysed as question-answer pairs, though within the constraints of a written, closed text, with no follow-up and no possibility of a challenge to the answer. As the FAQs are artificially constructed in such a way as to give the input for answers that inform, counterclaim, even perhaps counterattack, the work by Heritage (2002), Koshik (2003) and Léon (2004) is particularly helpful as they focussed on institutional, as opposed to conversational, questioning.

4. Analysis

Before looking at how expressions of space and time contribute to the construction of the identity and reputation of Unilever, a brief overview of how the FAQs themselves are constructed will give insight into the management of the business's reputation.

4.1. Question-answer format

The question-answer format is a 'watertight' device adopted by the company in order to safely manage its identity and reputation. Questions set up an agenda or topic for subsequent discussion, but can be worded in such a way as 'control' their answers through syntactic-semantic devices which reflect their illocutionary value (Diller quoted in Léon 2004: 1889). In fact, the FAQs present a range of questions which vary from neutral (open) questions simply seeking information to biased and more challenging questions, to simulate a variety of attitudes among the corporation's constituencies.

Three basic types of questions are employed: Wh- type questions, polar Yes/No questions and tag questions, and are distributed as follows: Wh-/Open 86, Yes/ No 39 and Tag 2. The most frequent form of question is the open type, which occurs with twice the frequency of the Yes/No type. An open type question is basically information-seeking, as for example:

(2) What do you expect the main challenges to be in order to achieve your objective?

(3) How will Unilever track progress against its nutrition targets?

However, open questions are not always as neutral as they would first appear to be. They can, in fact, become quite polemical, especially when combined with a Yes/No question. Both linguists and conversation analysts have addressed the issue of how some yes/no questions can be 'conducive' and 'prefer' an affirmative or negative answer over the other alternative (Heritage/Raymond 2012; Léon 2004). The preferred responses are those that align with the question, so grammatically affirmative questions expect affirmative answers and similarly grammatically negative questions expect negative answers. The preference or conduciveness is related to the design of the question or the assumption of the speaker. In the following example the first question is open and appears to be neutral and information seeking, but the second question challenges Unilever by throwing doubt on its efficiency and success in its campaigns.

(4) Who do you target with your campaigns? Are you reaching the people who might benefit most from changes in their hygiene habits?

Koshik suggests that 'reversed polarity questions', that is questions which suggest the opposite answer to what would be expected from their grammatical form, are asked from a strong or expert knowledge base which may help to explain how certain wh-questions are interpreted as implying assertions rather than as info-seeking questions and therefore can be seen as the content of ongoing disagreement, challenge or complaint (Koshik 2003: 72).

(5) How affordable are your products? Are you encouraging low income consumers to buy expensive international brands when they could buy a cheaper local product?

Challenges are also posed by the forefronting of the question with an initial statement which creates the context, but the question that follows is in contrast or conflict with the idea previously presented.

(6) Unilever's targets are heavily focused on addressing over-nutrition. What is Unilever doing to address the needs of poor people with specific nutritional deficiencies?

(7) Children will be the key audience for many of your campaigns. How does this fit with your commitments on marketing and advertising to children?

There are 15 questions in the FAQs with a negative interrogative form, such as those that start with *doesn't, isn't, can't.* Apart from expecting a negative answer, these questions are always strongly biased and therefore typically suggest some kind of contrast. They involve judgement, social evaluation and criticism on the part of the speaker and therefore act as assertions of opinions rather than questions. Here are some examples:

(8) Doesn't the power of your marketing work against sustainable consumption?

(9) With these hygiene initiatives, isn't Unilever's main objective just to sell more products?

(10) You have been producing sachets in developing countries for many years but are only now starting to look for solutions to the problems of sachet waste. Can't you accelerate your efforts in this area?

(11) Why don't you have a target to use 100% renewable materials in your packaging?

(12) Many companies have set 'zero waste to landfill' targets for their own manufacturing sites, but you're only committing to hold waste sent for disposal at 2008 levels. Why can't you do more?

(13) Hindustan Unilever is using women to sell its products. Isn't this exploitation of women?

All these forms set challenging questions which threaten the reputation of the company as they seem to undermine the values and mission of the company. Although there is a moral obligation to answer a question (Clayman 2001: 404), an answer may in fact be more or less evasive. Different strategies can be adopted to evade, such as responding only partially to the question and then developing the topic beyond or outside the parameters of the question or paraphrasing the question to modify it and facilitate the answer (Léon 2004). As there is no possible follow up in FAQs, the answer can be carefully orchestrated to steer the discussion in the direction the company wants and therefore to project a favourable image and defend its reputation. Indeed, there are few direct answers to the Yes/No questions in the FAQs. Of the 39 questions asked, only 9 actually get a direct answer, in the sense that there was a specific Yes/No answer, rather than just a response. Whilst Unilever would appear to courageously open itself up to accusations about its operations, its answers are managed to deflect the force of the argument and put forward a reasoned explanation and justification.

The following subsections will examine how space and time are used to construct this positive image and, above all, enhance its reputation.

4.2. Space

4.2.1. Space as location

Space can be conceived of in a number of real and figurative senses. In its most literal sense space is understood as location and in a globalised market a multinational cannot ignore local realities, local spaces. There are in fact numerous references to geographical areas, continents or countries, even names of cities, but as the following examples show they are in connection with sustainable projects and business activities seen from the perspective of Unilever rather than of the local communities.

(14) More than 30 tea estates have already achieved Rainforest Alliance certification, including Unilever's own estates in Kenya and Tanzania.

(15) As the Indian economy opened up in the early 1990s and competition increased, Hindustan Lever (now Hindustan Unilever) realised that in order to grow its business it needed to gain market access to the thousands of villages in rural India with under 2000 inhabitants. The company created a new business model called Project Shakti.

(16) To measure the impact of our 'Lifebuoy Way' handwashing education programme we conducted a clinical study in Mumbai between October 2007 and August 2008.

(17) The majority of Lifebuoy soap sales are in developing and emerging markets in Asia, Africa and Latin America. We target our activities in countries where the burden of relevant disease is highest and where typically there is also low and infrequent consumption of soap.

However, there are statements about the local benefits of these projects and strategies:

(18) It has helped double rural reach for Hindustan Unilever and it provides a means for many women in India to rise out of poverty and provide a better life and future for themselves and their families.

or where enormous benefits can be expected:

(19) We recently launched the Lifebuoy brand in Brazil where all our activities are focused in the north east of the country (Amazon area) where mortality and morbidity is highest.

4.2.2. Space as extension

There is in fact no specific identification of the company with a geographical place, even as its headquarters, and Unilever's truly multinational presence in all corners of the world is brought to the fore. In this sense, it is perhaps more correct to view space not as location but rather as extension, that is the area covered by the multinational's operations.

The company pushes back both physical and functional boundaries as it extends its activities. For example in (20) expansion can be interpreted as the physical growth of the company as indicated by the geographical names and numerical data reporting on the size of its operations and its sales.

(20) For example Lifebuoy soap has enjoyed double-digit growth over the last four years, with sales expanding from 20 countries in 2006, to 55 in 2010, reaching 1.1 billion consumers in Africa, Asia, Latin America, the Middle-East and Turkey.

But, perhaps even more importantly, expansion can reflect the extent of Unilever's impact on and its bearing in society. The following examples are based on the concordance lines for *across*, which was searched as indexical of the concept of extension. They show a broadening of the concerns of the company in all directions: *different income levels* 4, *the entire lifecycle* of its products (9,11-14,17,19) and *the value chain* (10,15 and 16), showing how the company views its responsibilities as going beyond the immediate business of manufacturing and selling. It also mentions specifically "the broad range of corporate responsibility issues" in 7.

(21)

1 perated an eco-efficiency manufacturing programme **across** all of its factory sites for over 15 years. The pr

2 the external announcement is an internal cascade **across** all our operations, so everyone with a role to pla

3 er these standards are not universally applicable **across**
all raw materials and geographies. For example the
4 markets? We sell branded food products for people **across**
different income levels all over the world and aff
5 'win', started in 2003 with 25 women in villages **across**
north Bangladesh. Joyeeta then became Project Apar
6 Programme we track changes in key nutrient levels **across**
our global foods portfolio and we report on the nu
7 t we have a long history of commitment and action **across**
the broad range of corporate responsibility issues
9 rom a single use, portion or serving of a product **across**
the entire lifecycle, including sourcing, manufact
10 s on track to meet the 2050 goal. Achieving this **across**
the entire value chain will be more challenging. O
11 20 target to halve GHG emissions per consumer use **across**
the lifecycle is a major step towards this long-te
12 now we must reduce the total environmental impact **across**
the lifecycle of our products. Unilever's manufac
13 to work towards with our suppliers and customers **across**
the lifecycle of our products. In our own operati
15 the absolute environmental impact of our products **across**
the value chain by a modest amount. The size of th
18 n allows us to do is to set comprehensive targets **across**
the whole organisation and focus the business on a
19 cs that deliver social and environmental benefits **across**
their entire lifecycle, while reducing dependency
20 . Each product category team has specific targets **across**
their product portfolio for water, waste and green
21 0, Telepresence will be available in 30 countries **across**
Unilever. By mid-2011 we will have installed 150 p

4.2.3. Space as ranking

Space can also be thought of in terms of ranking and Unilever does not hesitate to boast about its leading position in the field of sustainability. It proudly announces the external recognition awarded to the company for its efforts in sustainable agriculture (22) and also presents itself as a role model for others (23 and 24).

(22) Our Sustainable Agriculture programme has been in place for over 10 years, and in 2009 Unilever was named sustainability leader in the Food & Beverage sector of the Dow Jones Sustainability Index for the eleventh consecutive year.

(23) Where we have influence in the world market in terms of how much we buy, we will continue to use this wisely and lead by example (such as tea and palm oil). Where we are relatively small players, such as for dairy and sugar, we will partner with other organisations to drive the agenda.

(24) While our ambition remains to reduce accidents to nil, our current performance places us among the leading companies in our sector in terms of safety.

In addition, common spatial indicators such as *here* do not refer to location, but rather an area of the business:

(25) It covers the full value chain from sourcing through to consumer use and disposal. Even **here**, by achieving this target we will reduce the absolute environmental impact of our products across the value chain by a modest amount.

4.2.4. Space as expertise

As all these examples suggest, space is, in fact, defined in terms of business, as a professional space, which the multinational occupies and dominates. The discourse focuses entirely on what Unilever is doing. This can also be seen by the use of *we* and *our*. All but one of the 322 occurrences of *we* are exclusive and that exception is in a question. The most frequent collocates of *our* are *products* (19 occurrences), *campaigns* (11), *business* (10) and *suppliers* (10). Coupland and Brown's view of identity construction (2004) as an insider-outsider dialogue, therefore, is a very apt description of the spatial framework in the website, though the dialogue must inevitably be one-sided in FAQs.

The professional space carved out by Unilever covers three main areas, firstly, its sales and the extent of its activities as we have seen above, secondly, the expertise of the business and thirdly, the socially responsible roles it professes to play. The expertise of the company is projected by emphasising the technical aspects of the topics.

(26) During 2009 we conducted a study in Asia to test new technology options. This identified pyrolysis as a promising opportunity. Pyrolysis turns sachet waste into fuel, recovering around 60% of the embedded energy in the sachet. We are piloting this and similar energy recovery technologies in four countries to assess their technical and commercial feasibility.

Twelve questions actually use the What + verb sequence prompting Unilever to clarify certain concepts, with questions like *What do you mean by...?* or *What is the definition of...?* and thus highlight the company's knowledge and professionalism on topics such as *recovery* as opposed to recycling in waste management, 'per consumer use' and 'cradle to cradle'.

(27)

```
1       n of self-verification and performance reporting. What do
you mean by 'certified' sustainable sources? We
2       ssess their technical and commercial feasibility. What do
you mean by 'cradle to cradle'? The 'cradle to
3       n the waste packaging is one way to resolve this. What do
you mean by 'recovery'? Recovery refers to a va
4      t a social as well as an environmental assessment. What do
you mean by 'water extraction'? Water is a comp
8       o how consumers use and dispose of our products.  What does
Unilever mean by 'product lifecycle'? The tot
9      istribution to consumer use and eventual disposal. What does
Unilever mean by the phrase 'per consumer use
11     when they wash their clothes or clean their teeth. What is
the definition of a 'water-scarce country'? Unl
12      t in our Policy's requirements we will accept it. What is
Unilever's definition of "recycled" material? F
```

It also stresses its expertise through numerous references to partnerships and alliances with universities and leading experts in the fields of health, nutrition, waste management, and other areas of sustainability.

(28) How do you propose to drive up recycling rates? People have been trying to do this for years. What difference can you make?
We must proactively facilitate collection, recycling and recovery by providing technical assistance, helping to shape policies, providing access to finance and market guarantees. To do this we have to work together with NGOs, governments and the private sector.

4.2.5. Space as social commitment

In the following extract Unilever presents the roles it envisages for itself in society, which "blur the boundaries between the role of organisations and the role of governments [...] increasingly assuming roles that were once regarded as belonging to the public sector" (Tregidga *et al* 2008: 15). The example shows how Unilever manages to turn around a challenging negative interrogative into a well-argued defence of the company's position and reputation and at the same time project a very positive image of its professional space, reflecting the identity and the roles it envisages for itself:

(29) Doesn't the power of your marketing work against sustainable consumption?
Advertising and marketing can be a powerful force for behaviour change and, used effectively, allows us to encourage consumers to adopt healthy habits, such as brushing teeth twice a day, as well as engage with them on sustainability issues, such as reducing energy and water use when washing their clothes.
At the same time we recognise that there are concerns about the role of marketing in encouraging what can be seen as unnecessary and unsustainable consumption.
A large proportion of Unilever's future growth will come from developing markets where a whole segment of the population is emerging from the bottom of the economic pyramid to become first-time buyers of products such as soap, toothpaste and shampoo.
Our portfolio of products meets basic everyday needs for nutrition, hygiene and personal care. This gives us confidence that we are not encouraging irresponsible and unnecessary consumption, but rather helping to improve people's quality of life.
In addition, branded products like ours guarantee consistent quality. This is just as important for people on limited incomes, who cannot afford to risk money on unreliable products, as it is for more affluent consumers.
However we do need to do our utmost to encourage more sustainable patterns of consumption in developing markets and our innovation efforts are now increasingly geared towards this challenge.

Unilever responds to the threatening question by picking up on the words *power* and *against* and turns the negative bias of the question into the positive phrase *a powerful force for behaviour change*. And it then goes on to carve out for Unilever a very broad sphere of influence in society and the world, which extends far beyond the manufacturing side of the business. Unilever outlines the roles it plays and how it uses the very tools of its trade – advertising and marketing – to benefit society as a whole, thus presenting itself as an actor in the social world, legitimizing its extensive business activities (Breeze 2012: 4).

Unilever presents itself, firstly, as an 'educator' through its campaigns to improve hygiene and to encourage a more responsible attitude to environmental issues: *allows us to encourage consumers to adopt healthy habits* and *engage with them on sustainability issues.* Secondly, it presents itself as a 'provider' of products essential to society which not only satisfy the needs of people, but also help to improve their quality of life. This positions the organization not on the edge of society as just a producer, but as fulfilling fundamental social functions which benefit whole communities. Thirdly, it presents itself as a 'responsible' business producing quality products to avoid wasting money, especially for low-income consumers. It is interesting to note that these roles and characteristics are typical of the image companies generally want to project (Breeze 2012; Tregidga *et al.* 2008).

Unilever also takes on the role of 'empowering' people, whether they be consumers, smallholders or others, in social, economic and environmental questions:

(30) This means working with our suppliers as well as offering new and innovative products to consumers that can help them reduce their own impact on the environment.

(31) However, 85% of the world's farmers are smallholders, farming less than two hectares. These smallholders often harvest much lower crop yields than large-scale producers, because they lack access to basic services: good varieties of crops and seeds, knowhow, agricultural inputs such as fertilisers, storage facilities and routes to market. Providing access to these services can double, triple or even quadruple their yields within one or two years.

A company's reputation relies heavily on the trust of stakeholders and therefore its relationship with 'outsiders' is fundamental, thus creating a real need to establish a good rapport with them. Although Unilever portrays itself as taking a proactive and constructive role in society, both in social and environmental issues, it always maintains a clear demarcation line between 'insider' and 'outsider' and emphasises its authoritative position. In fact, Unilever makes the distinction between insiders and outsiders explicit in the expressions *inside our factory gates* and *within and outside our company* in the following examples:

(32) Therefore, if we wish to have a sustainable business and society we must work across the whole lifecycle and not focus solely on activities *inside our factory gates*. This means working with our suppliers as well as offering new and innovative products to consumers that can help them reduce their own impact on the environment.

(33) Can you ever source 100% of your agricultural raw materials sustainably?
Yes, we believe we can. We know it will take a huge amount of effort due to the scale and complexity of the supply chains we are operating in. But the 100% target has been set to change behaviour *within and outside our company* and it is starting to do this.

4.3. Time

As described in the introduction, reputation is built up over time and based on what has been achieved so far. In contrast, sustainability seems to consist in a series of promises which can be measured and verified only in the future. These two contrasting perspectives are reconciled through a constant alternation and interplay between past, present and future tenses in the texts. The past signals the beginning of studies, analysis, planning and projects to prepare the ground for the implementation of the sustainability plan, the present to describe how the plan works and what is being done and the future to mark out the targets and phases of the plan as milestones and objectives. The following extract illustrates the interplay of tenses.

(34) What makes the Unilever Sustainable Living Plan different from those your competitors are publishing?
Each company is different, and is at different stages of its journey to understanding the implications of a commitment to sustainable development. We have concentrated on the areas where our business can make the biggest impact, given the range of our brand portfolio in food, home and personal care. We have taken into account where we operate now and where we hope to grow.
We think our plan is differentiated in several ways. Recognising the scale of challenges all companies face, our plan explicitly addresses the complete value chain from suppliers through production to distribution and final use. It covers environmental, social and economic aspects of sustainability. It is built around clear commitments by specific dates. Underneath our three headline

> commitments we have over 50 publicly-stated goals with target dates for completion. We will report our progress every year.
> We also think our plan is unique in terms of the rigour of the analysis we carried out to set appropriate targets. We conducted an extensive baseline measurement of our global product portfolio to prioritise where to set targets. We developed metrics for greenhouse gases, water and waste, taking a lifecycle approach. We then assessed over 1 600 products in 14 countries using these metrics.
> We are fortunate in being able to base this work on more than 15 years' work on eco-efficiency in manufacturing, 10 years on sustainable agricultural sourcing, and a company-wide ethos that stretches back to our founders' sense of social purpose more than 100 years ago.

This particular text focuses on how the plan has been conceived and its implementation. The present tense is used to give information about the plan and its rationale (*addresses*, *covers*, *is built*, *have*), whereas the experience, skills and knowledge acquired so far by the company is expressed through the past tense (*developed*, *conducted*, *assessed*). The present perfect tense acts as a link between the past and the present (*we have concentrated*, *we have taken into account*) to reinforce the idea of the present work being based on many years of experience. The final sentence draws the reader's attention to the ethically driven nature of Unilever's activities. Although it is not actually defined or described, an ethos has always acted as the guiding principle of the company from its very origins many, many years ago. It is interesting to note that there is no explicit evaluative language, either in the form of superlatives or highly positive expressions. The thorough and rigorous analyses carried out by the company and the rationale behind the plan convey implicitly, and perhaps more convincingly, the professionalism of the company and therefore the good reputation of the company.

The next example also builds its argument on past experience and successes, *Unilever brands have real experience*, which is now producing *formidable* results. Together the past and the present provide the basis for future progress.

(35) We have set nutritional targets, based on very strict global dietary guidelines. Currently around 20% of our portfolio meets these stringent targets. We aim to double that to 40% by 2020. This will be a formidable achievement, given

> that these changes come on top of five years' work to improve our products. During this time we have implemented many of the biggest opportunities for improvement.
> Through the *expertise* we bring from our marketing, Unilever brands have real experience in influencing consumer behaviour – and can show *past successes* on health and hygiene through handwashing, oral care and cholesterol reduction, for example. Our Cleaner Planet Plan *is succeeding in engaging* consumers on environmental issues too.

Example (36) asks a very simple question, but the answer explains in detail how the present plan is devised to give results within a relatively limited period of time, though only as a stepping stone towards new, more ambitious targets in the future, thus conveying the idea of continuity.

(36) What happens after 2020?
We have focused on the year 2020 as we want our plan to be concrete and to deliver within a meaningful time frame. We recognise this is only a milestone on a longer journey. We will continuously update the plan and as we get closer to 2020 we will revisit the plan, its goals and targets and re-set these for the next time horizon.

Table 2 shows how the different tenses are distributed in the FAQs for verbs used with *we*.

	Past		Present		Future	
WE	Present perfect	42	Present simple	189	will	43
	Past simple	24	Present continuous	22	going to	1

Table 2. Distribution of tenses.

There is in fact a strong preponderance of the present tenses indicating the plan is actually being implemented and is well underway. It is interesting to notice the use of *will* rather than *going to* for the future. There is just one occurrence of *going to* as opposed to the 43 occurrences of *will*, which indicates the firm intention of Unilever to implement its plan, but also its moral commitment to the ideals it has set itself. *Will* performs a kind of promise to its stakeholders and the public at large.

Although the number of occurrences for the future tenses is smaller than for the past and especially the present, there are other temporal indicators which point to a relatively strong emphasis on the future in the FAQs. For example, when the text was run through WMatrix[3] the third key semantic domain was 'wanted', which included words like *target*, *plan*, *campaign*, *programme*, *aim*, *objectives*, *choice*, *ambition*, *wish* and so on, positioning the work of the company in this field as a long term endeavour.

Indeed, the central and uniting temporal theme of the FAQs is progression, as highlighted in the following extract.

(37) Are you on track to meet the targets you have set?
Yes. We are already world leaders in the sustainable sourcing of palm oil and tea. In 2009, we purchased 185 000 tonnes of certified sustainable palm oil via Green Palm Certificates, accounting for 15% of our total purchases. This is enough to cover our requirements in Europe, Australia and New Zealand. In 2010 we will double this to 30% and double it again to 60% in 2011. We shall continue to do this until we reach our 100% target by 2015.

The transformation of a company into an environmentally friendly and socially sustainable business requires extensive organization and effort. By emphasizing the 'work in progress' and the numerous targets set inside the company and not imposed from outside authorities, Unilever tries to deflect some of the heavy criticisms laid against multinationals about not doing enough now, as suggested in some of the very challenging questions (8-13) in Section 4.1. The insistence on the numerous objectives the company is fixing is shown by the very frequent use (38 in all) of the phrase composed of the preposition *by* + year, as for example *by 2020*, *by 2015* and even the very distant *by 2050* as in (38).

(38) In our own operations, we are confident that despite growing as a business we will achieve absolute reductions in greenhouse gases. In particular we will

---

3 A software developed by Lancaster University (Rayson 2005) which includes a semantic tagger "to assign a pre-programmed meaning category to every word [...] The program then compares the frequencies of individual categories to [...] the British National Corpus. The 'keyness' (i.e. statistically significant over-use or under-use) of a particular semantic domain in the texts under scrutiny is established using a log-likelihood test" (Koller 2007: 117).

> meet the United Nations' requirement of reducing GHGs by 50-85% by 2050 in order to limit global temperature rise to two degrees.

The idea of progression is also reinforced by a number of adjectives and adverbs of manner or time which individually do not have a high frequency, but if grouped together represent a broad semantic domain and have a significant number of occurrences: *constant(ly)* 2, *continuous* 4, *continuously* 4, *continuing* 1, *ongoing* 3, *sustained* 2. Other expressions include *already* 12, *now* 8, *future* 7, *past* 4 and *ahead* 3, which once again suggest the continuity of time.

5. Concluding remarks

The analysis of the spatial and temporal frameworks in the FAQs has conveyed an image of the company as big, powerful and supranational, but also as a 'caring', socially involved organization, which combines, unashamedly and presumably to the advantage of all, its profit maximizing strategies with both short and long term social and environmental objectives. In essence, it projects the identity of a 'Global Benefactor' (Breeze 2012: 7).

This image is conveyed primarily by the spatial expressions which offer different perspectives on the business's activities, the range of projects and sphere of influence. In contrast, though as to be expected, time added the dimension of reputation to the corporate image. Time is portrayed as a continuum, stretching back in time and forward into a fairly distant future, but also broken down into milestones and targets.

A corporate website does in fact create an ideological space within which an institution can operate, to justify and legitimize its activities by responding to and pre-empting criticism in order to enjoy sufficient social acceptance to pursue its activities freely (Breeze 2012: 4). The FAQs format provides a safety net in which to tackle the most challenging topics. Indeed, rather than repudiating accusations, Unilever actually incorporates the 'negative' aspects into their core values and turn them into virtues, as the size, growth and business of the company will bring far-reaching changes to the world all over.

Space in computer mediated communication can be described as virtual and unbounded, but at the same time we talk about websites as containers, as the expression "in the website" clearly shows. This metaphor is reinforced by the 'exclusive' use of the exclusive *we* to indicate the actions and work of the company in its professional space. No attempt is made to reach out to the reader and the 'insider' position dominates the text. Space in all its various conceptual interpretations provides insight into the many facets of the corporate identity, its activities, its purposes, its roles, whereas time highlights the image of professionalism, its expertise and pioneering work in the field of sustainability, but primarily in terms of reputation.

There is, however, a sense of placelessness in the website and a kind of omnipresence of the company in the world. Indeed, placelessness appears to be a common feature of some virtual genres. As Myers says about blogs, location is not their defining characteristic, but rather what the person is doing (2010: 49). Although obviously unable to enjoy the same freedom as a personal blog within the institutional context of a corporate website, space and time references are used in the FAQs, as in blogs (Myers 2010: 9), for the specific rhetorical purposes of contrasting, justifying and evaluating. Unilever contrasts what has been achieved already and what will be done, it justifies and legitimizes its expansionary strategies and positively evaluates its ethical approach and operations that improve the lives of a wide range of beneficiaries.

If we just briefly recall Ford's comment quoted in the introduction about reputation being built on what has been achieved, the results of the analysis would seem to challenge this statement. The focus is on the present, with a strong leaning for the future, suggesting that Unilever is emphasizing its present commitment to sustainability as a continuation of work already underway, but also, perhaps, to safeguard itself against any failures or shortcomings which could damage its reputation in the future.

References

Abdullah, Zulhamri 2009. Beyond Corporate Image: Projecting International Reputation Management as a New Theoretical Approach in a Transitional Country. *The Journal of Economics and Management* 3/1, 170-183.

Balmer, John M.T. / Fukuwawa, Kyoko / Gray, Edmund R. 2007. The Nature and Management of Ethical Corporate Identity: A Commentary on Corporate Identity, Corporate Social Responsibility and Ethics. *Journal of Business Ethics* 76/1, 7-15.

Benwell, Bethan / Stokoe, Elizabeth 2006. *Discourse and Identity*. Edinburgh: Edinburgh University Press.

Breeze, Ruth 2012. Legitimation in Corporate Discourse: Oil Corporations after Deepwater Horizon. *Discourse and Society* 23/1, 3-18.

Clayman, Steven 2001. Answers and Evasions. *Language in Society* 30/3, 403-442.

Coupland, Christine / Brown, Andrew 2004. Constructing Organizational Identities on the Web: A Case Study of Royal Dutch/Shell. *Journal of Management Studies* 41/8, 1325-1347.

Galton, Antony 2011. Times flies but space does not: limits to the spatialisation of time. *Journal of Pragmatics* 43/3, 695-703.

Gray, Edmund R. / Balmer, John M.T. 1998. Managing Corporate Image and Corporate Reputation. *Long Range Planning* 31/5, 695-702.

Heritage, John 2002. The limits of questioning: negative interrogatives and hostile question content. *Journal of Pragmatics* 34/10-11, 1427-1446.

Heritage, John / Raymond, Geoffrey 2012. Navigating Epistemic Landscapes: Acquiescence, Agency and Resistance in Responses to Polar Questions. In de Ruiter, Jan P. (ed.) *Questions*. Cambridge: Cambridge University Press, 179-192 .

Koller, Veronika 2007. 'The World's Local Bank': Glocalization as a Strategy in Corporate Branding Discourse. *Social Semiotics* 17/1, 111-131.

Koshik, Irene 2003. Wh-questions as challenges. *Discourse Studies* 5, 51-77.

Léon, Jacqueline 2004. Preference and "bias" in the format of French news interviews: the semantic analysis of question-answer pairs in conversation. *Journal of Pragmatics* 36/10, 1885-1920.

Löwensberg, David 2006. Corporate image, reputation and identity. In Trench, Ralph / Yeoman, Liz (eds) *Exploring Public Relations*. Harlow: FT/Prentice Hall, 237-251.

Myers, Greg 2010. *The Discourse of Blogs and Wikis*. London: Continuum.

Pruzan, Peter 2001. Corporate Reputation: Image and Identity. *Corporate Reputation Review* 4/1, 50-64.

Shinkle, George A. / Spencer, J. William 2012. The social construction of global corporate citizenship: Sustainability reports of automotive corporations. *Journal of World Business* 47/1, 123-133

Tregidga, Helen / Milne, Marcus / Kearins, Kate 2008. *(Re)Presenting Sustainable Organizations: A New Discursive Identity*. Anaheim, CA, USA: 2008 Academy of Management Annual Meeting, 8-13 Aug 2008, conference paper available at <www.acis.canterbury.ac.nz/documents/Markus_Milne_ResearchProgrammes/Re_Presenting_Sustainable_Organizations.pdf>, downloaded May 17, 2011.

Turnbull, Judith 2011. How 'glocal' is corporate discourse? A case study of a multinational's website. In Salvi, Rita / Tanaka, Hiro (eds) *Intercultural Interactions in Business and Management*. Bern: Peter Lang, 73-90.

Van Riel, Cees 1995. *Principles of Corporate Communication*. Hemel Hempstead: Prentice Hall.

Notes on Contributors

GIUSEPPE BALIRANO, PhD in English for Special Purposes, is senior lecturer of English language and translation studies at the University of Naples 'L'Orientale'. He is currently lecturing English linguistics in the MA courses in 'Language and Intercultural communication in the Euromediterranean area'. His most recent research interests lie in the field of CDA in a multimodal perspective, humour in cross-cultural interaction and audiovisual translation.

JANET BOWKER is a tenured researcher and Adjunct Professor in English Linguistics and Translation in the Faculty of Economics, 'Sapienza' University of Rome. Her research interests lie mainly in academic and professional discourse analysis, corpus linguistics, and socio-cognitive pragmatics. Her self-compiled specialised corpus of business discourse has investigated a variety of topics including the processes of description, evaluation and persuasion in internal and intercultural corporate discourse, and multimodality, genre and register in virtual professional 'space'. Among her latest publications are 'The language of numerical data and figures in internal company communications: the processes of subjectivization and relevance-making', in *International Journal of Applied Linguistics* (2011) and 'The creation of trust in intercultural business consulting encounters: global corporate practices in local contexts' in *Discourse and Trust: Perspectives from Text and Talk in Organizations*, G. Jacobs, K. Pelsmaekers, C. Rollo (eds), Amsterdam: John Benjamins (forthcoming).

GIUDITTA CALIENDO is Researcher and Adjunct Professor in English Linguistics and Translation at the University of Napoli 'Federico II', Italy. She has been a Fulbright Research Scholar at the EU Center of Excellence of the University of Washington, USA. Her research interests include political and institutional discourse, critical discourse analysis, knowledge dissemination, specialized translation and language policy in international organizations. She has published on the above areas both in edited volumes (*Linguistic Insights*, Bern: Peter Lang) and in refereed journals (*Journal of Contemporary European Research*; *Journal of the American Translation and Interpreting Studies Association*). She is co-editor of the volumes: *Genre(s) on the move: Hybridization and Discourse Change in Specialized Communication* (with S. Sarangi and V. Polese, ESI, 2011); *The Language of Popularization. Theoretical and Descriptive Models* (with G. Bongo, Peter Lang, forthcoming).

GIULIANA DIANI is a tenured researcher in English Language and Translation at the University of Modena and Reggio Emilia. She holds an MA in Language Studies from the University of Lancaster (UK) and a PhD in English Linguistics from the University of Pisa (Italy). She has worked on various aspects of discourse analysis and EAP, with special reference to argumentation, metadiscourse and evaluative language. Her recent work centres on language variation across academic genres, disciplines and cultures through the analysis of small specialized corpora. Her recent publications include: *Reviewing academic research in the disciplines: insights into the book review article in English* (2012); *La trasmissione del sapere nelle diverse comunità accademiche: una prospettiva plurilingue* (co-edited with Chiara Preite, 2012); *Academic evaluation: review genres in university settings* (co-edited with Ken Hyland, 2009).

LAURA FERRAROTTI is a researcher of English Language and Linguistics in the Department of Communication and Social Research of the University of Rome 'Sapienza', where she teaches Business English and English language. Her research interests span from corpus linguistics and discourse analysis to mass media studies, from the social sciences to promotional discourse and advertising language. She has recently published on TV news broadcasting and on advertising. Her current research deals with the linguistic features of Italian university websites in a comparative perspective.

ERSILIA INCELLI is an English language lecturer at the Faculty of Economics, 'Sapienza' University of Rome. Her current research interests are in socio-linguistics, discourse analysis and corpus linguistics, with particular reference to professional identities and the workplace. Recent research has focused on combining these approaches to analyse specialized genre, especially in business and legal discourse, to show how quantitative and qualitative techniques of analysis can aid language enquiry and sociolinguistic research, for example, analysing metaphors in immigration law, or identifying specific lexical patterns which characterize a business community. Her teaching activities involve language development for academic and professional purposes.

DONATELLA MALAVASI holds a Ph.D. in Comparative Languages and Cultures from the University of Modena and Reggio Emilia, where she is now a Tenured Researcher in English Linguistics and Translation. Her research interests span from corpus linguistics and translation to genre and discourse analysis. Her most recent contributions focus on the analysis of financial communication and Corporate Social Responsibility Reports ('The Multifaceted Nature of Banks' Annual Reports as Informative, Promotional and Corporate Communication Practices', 2010; "'Doing Well by Doing Good': A Comparative Analysis of Nokia's and Ericsson's Corporate Social Responsibility Reports", 2011; "'The Necessary Balance between Sustainability and Economic Success': an Analysis of Fiat's and Toyota's Corporate Social Responsibility Reports", 2012), the construction of identity in academic discourse ("'I hope that I can convince readers of…': the construction of professional identity in marketing research articles", 2009), and the cross-disciplinary study of keywords ('History v. Marketing. Keywords as a clue to disciplinary epistemology', 2010; with D. Mazzi).

GILLIAN MANSFIELD is Associate Professor of English language and translation at the University of Parma. Her present research interests include wordplay in media texts, cultural identity in global advertising and the popularizing features of texts in museum websites. Her most recent publications include: 'And Murder She Wrote: un'indagine (quasi poliziesca) fra i titoli dei romanzi del mistero di scrittrici anglo-americane' (2011) and 'Laughter in the lecture theatre – a serious matter. Studying verbal humour in the British TV sitcom' (2012). She has been both treasurer and secretary of AICLU (Associazione Italiana dei Centri Linguistici Universitari). From September 2008 to September 2012 she became Secretary General of CercleS (European Confederation of Language Centres in Higher Education) after which she was elected President for the period 2012-14. In her capacity as President she has also become, together with David Little, Editor-in-Chief of the CercleS academic journal *Language Learning in Higher Education.*

MARIA CRISTINA NISCO holds a PhD in Cultural and Postcolonial Studies of the Anglophone Countries, she is currently adjunct professor in English language at the University of Molise and the University of Naples 'L'Orientale'. She has recently started her second PhD in English for Special Purposes at the University of Naples 'Federico II'. She has researched and published on postcolonial theory and diasporic literature, visuality and multimodal discourse, varieties of English and translation.

FRANCA POPPI is Associate Professor of English Language and Translation at the University of Modena and Reggio Emilia, where she is also Director of the Master Course in Languages for Communicating in Businesses and International Organizations. Her initial work was concerned with learner autonomy and advising in self-instruction, but she has since then concentrated on the linguistic, social and psychological dimensions of discourse analysis. Her research interests have focused on the interactional features of discourse, with particular reference to academic settings (economics and marketing textbooks) and the language of the law. Her current research areas include, beside discourse analysis and language variation, intercultural communication and the sociolinguistics of English as a lingua franca.

CHIARA PROSPERI PORTA has been working at 'Sapienza' University of Rome since 2002. Throughout this time she has taught at B.A. and M.A. levels in the Faculties of Economics, Humanities and Law. She has also been assistant researcher of English in the Faculty of Economics and has been teaching English for tourism at the ITS Foundation in Rome. Her interests are mainly in ESP and EAP with particular reference to scientific discourse and genre analysis, technological genre and communication in professional fields. She has published several articles on these subjects. Some of her recent essays are: 'Analysing discourse in research genre: the case of biostatistics' (John Benjamins 2013 forthcoming); 'The metaphor of illness and health in economics'(Edizioni Q, Roma 2011); 'Bringing the world to the classroom: websites and what they can teach' (Bulzoni, Roma 2010).

RITA SALVI is full professor in English Linguistics and Translation in the Faculty of Economics, 'Sapienza' University of Rome. She is a member of AIA (Italian Association of English Studies) and former secretary and treasurer (2005 – 2009); member of *Eurolinguistics* Association and AIDEL (Italian Association of Law and Literature). Representative of 'Sapienza' in the Directive Board of *CLAVIER, Corpus and Language Variation in English Research*, in collaboration with the universities of Modena-Reggio Emilia, Bergamo, Firenze and Siena. Member of the Coordinators' Committee of the Seminar 'Lionello R. Levi Sandri', a permanent research group on transnational law and legal translation studies. Her research interests include general and applied linguistics, text and genre analysis, intercultural discourse for academic and professional purposes, language teaching particularly through the medium of advanced technologies, synchronic and diachronic lexicographic analysis mainly in the field of specialized language (economics and law). She serves as reviewer for journals, book publishers and research funding entities. Member of the Editorial Board of the journals *International Journal of Law, Language and Interdisciplinary Practice* and *American Journal of Linguistics*, and of the linguistics series *Archives*. She sits on various teaching and research committees, as well as in scientific committees of several national and international conventions. She is co-editor (with Hiromasa Tanaka) of the volume *Intercultural Interactions in Business and Management*, Bern: Peter Lang, 2011. Her recent publications include: 'Transfer of knowledge and academic discourse: lectures on economics' (2012); 'Textual equivalence in the translation of specialized languages: discourse adherences and dissimilarities' (2012); 'Il linguaggio legale: dalla parola al testo' (2011); 'Appraisal Theory as a methodological proposal for stylistic analysis' (2010).

CHRISTINA SAMSON (Ph.D.) is an EAP/ESP lecturer in the University of Florence. Her research interests cover textual language variation in specialised corpora; corpus linguistics analysis of accounting narratives and websites; language-conceptual transfer strategies between L1 and EFL in students at tertiary level. She has published in peer reviewed international volumes/journals and has served as referee. Her publications include: *Comprehending Domain-Specific Texts: A Contextual Syncretism Approach* (2012); 'From cultural islands to popular sites. Semantic sequences typifying museum descriptions on the Web' (2012); 'Ex-sacred territories on the Internet. Examples of space, identity and discourse interconnectedness in museum websites'(2011); 'Trust the text. A corpus-driven analysis of accounting narratives' (in press).

JUDITH TURNBULL is a foreign-language assistant at the Faculty of Economics, 'Sapienza' University of Rome. Her research interests include discourse analysis applied to the fields of economics and law and cross-cultural business communication, with a particular focus on websites. Her recent publications include: 'The Blogs of the European Commission: Public-private Space in a Professional Institutional Context' (2013); 'How 'Glocal' is Corporate Discourse? A Case Study of a Multinational's Website' (2011); 'Harmonisation of the law and legal cultures in the EU: A Linguistic Approach' (2010); '(Self-) Description and Evaluation in Corporate Websites' (2009).

Linguistic Insights

Studies in Language and Communication

This series aims to promote specialist language studies in the fields of linguistic theory and applied linguistics, by publishing volumes that focus on specific aspects of language use in one or several languages and provide valuable insights into language and communication research. A cross-disciplinary approach is favoured and most European languages are accepted.

The series includes two types of books:

- **Monographs** – featuring in-depth studies on special aspects of language theory, language analysis or language teaching.
- **Collected papers** – assembling papers from workshops, conferences or symposia.

Each volume of the series is subjected to a double peer-reviewing process.

Vol. 1 Maurizio Gotti & Marina Dossena (eds)
Modality in Specialized Texts. Selected Papers of the 1st CERLIS Conference.
421 pages. 2001. ISBN 3-906767-10-8 · US-ISBN 0-8204-5340-4

Vol. 2 Giuseppina Cortese & Philip Riley (eds)
Domain-specific English. Textual Practices across Communities and Classrooms.
420 pages. 2002. ISBN 3-906768-98-8 · US-ISBN 0-8204-5884-8

Vol. 3 Maurizio Gotti, Dorothee Heller & Marina Dossena (eds)
Conflict and Negotiation in Specialized Texts. Selected Papers of the 2nd CERLIS Conference.
470 pages. 2002. ISBN 3-906769-12-7 · US-ISBN 0-8204-5887-2

Vol. 4 Maurizio Gotti, Marina Dossena, Richard Dury, Roberta Facchinetti & Maria Lima
Variation in Central Modals. A Repertoire of Forms and Types of Usage in Middle English and Early Modern English.
364 pages. 2002. ISBN 3-906769-84-4 · US-ISBN 0-8204-5898-8

Editorial address:

Prof. Maurizio Gotti — Università di Bergamo, Facoltà di Lingue e Letterature Straniere,
Via Salvecchio 19, 24129 Bergamo, Italy
Fax: 0039 035 2052789, E-Mail: m.gotti@unibg.it

Vol. 5 Stefania Nuccorini (ed.)
Phrases and Phraseology. Data and Descriptions.
187 pages. 2002. ISBN 3-906770-08-7 · US-ISBN 0-8204-5933-X

Vol. 6 Vijay Bhatia, Christopher N. Candlin & Maurizio Gotti (eds)
Legal Discourse in Multilingual and Multicultural Contexts.
Arbitration Texts in Europe.
385 pages. 2003. ISBN 3-906770-85-0 · US-ISBN 0-8204-6254-3

Vol. 7 Marina Dossena & Charles Jones (eds)
Insights into Late Modern English. 2nd edition.
378 pages. 2003, 2007.
ISBN 978-3-03911-257-9 · US-ISBN 978-0-8204-8927-8

Vol. 8 Maurizio Gotti
Specialized Discourse. Linguistic Features and Changing Conventions.
351 pages. 2003, 2005.
ISBN 3-03910-606-6 · US-ISBN 0-8204-7000-7

Vol. 9 Alan Partington, John Morley & Louann Haarman (eds)
Corpora and Discourse.
420 pages. 2004. ISBN 3-03910-026-2 · US-ISBN 0-8204-6262-4

Vol. 10 Martina Möllering
The Acquisition of German Modal Particles. A Corpus-Based Approach.
290 pages. 2004. ISBN 3-03910-043-2 · US-ISBN 0-8204-6273-X

Vol. 11 David Hart (ed.)
English Modality in Context. Diachronic Perspectives.
261 pages. 2003. ISBN 3-03910-046-7 · US-ISBN 0-8204-6852-5

Vol. 12 Wendy Swanson
Modes of Co-reference as an Indicator of Genre.
430 pages. 2003. ISBN 3-03910-052-1 · US-ISBN 0-8204-6855-X

Vol. 13 Gina Poncini
Discursive Strategies in Multicultural Business Meetings.
2nd edition. 338 pages. 2004, 2007.
ISBN 978-3-03911-296-8 · US-ISBN 978-0-8204-8937-7

Vol. 14 Christopher N. Candlin & Maurizio Gotti (eds)
Intercultural Aspects of Specialized Communication.
2nd edition. 369 pages. 2004, 2007.
ISBN 978-3-03911-258-6 · US-ISBN 978-0-8204-8926-1

Vol. 15 Gabriella Del Lungo Camiciotti & Elena Tognini Bonelli (eds)
Academic Discourse. New Insights into Evaluation.
234 pages. 2004. ISBN 3-03910-353-9 · US-ISBN 0-8204-7016-3

Vol. 16 Marina Dossena & Roger Lass (eds)
Methods and Data in English Historical Dialectology.
405 pages. 2004. ISBN 3-03910-362-8 · US-ISBN 0-8204-7018-X

Vol. 17 Judy Noguchi
The Science Review Article. An Opportune Genre in the Construction of Science.
274 pages. 2006. ISBN 3-03910-426-8 · US-ISBN 0-8204-7034-1

Vol. 18 Giuseppina Cortese & Anna Duszak (eds)
Identity, Community, Discourse. English in Intercultural Settings.
495 pages. 2005. ISBN 3-03910-632-5 · US-ISBN 0-8204-7163-1

Vol. 19 Anna Trosborg & Poul Erik Flyvholm Jørgensen (eds)
Business Discourse. Texts and Contexts.
250 pages. 2005. ISBN 3-03910-606-6 · US-ISBN 0-8204-7000-7

Vol. 20 Christopher Williams
Tradition and Change in Legal English. Verbal Constructions in Prescriptive Texts.
2nd revised edition. 216 pages. 2005, 2007. ISBN 978-3-03911-444-3.

Vol. 21 Katarzyna Dziubalska-Kolaczyk & Joanna Przedlacka (eds)
English Pronunciation Models: A Changing Scene.
2nd edition. 476 pages. 2005, 2008. ISBN 978-3-03911-682-9.

Vol. 22 Christián Abello-Contesse, Rubén Chacón-Beltrán, M. Dolores López-Jiménez & M. Mar Torreblanca-López (eds)
Age in L2 Acquisition and Teaching.
214 pages. 2006. ISBN 3-03910-668-6 · US-ISBN 0-8204-7174-7

Vol. 23 Vijay K. Bhatia, Maurizio Gotti, Jan Engberg & Dorothee Heller (eds)
Vagueness in Normative Texts.
474 pages. 2005. ISBN 3-03910-653-8 · US-ISBN 0-8204-7169-0

Vol. 24 Paul Gillaerts & Maurizio Gotti (eds)
Genre Variation in Business Letters. 2nd printing.
407 pages. 2008. ISBN 978-3-03911-681-2.

Vol. 25 Ana María Hornero, María José Luzón & Silvia Murillo (eds)
Corpus Linguistics. Applications for the Study of English.
2nd printing. 526 pages. 2006, 2008. ISBN 978-3-03911-726-0

Vol. 26 J. Lachlan Mackenzie & María de los Ángeles Gómez-González (eds)
Studies in Functional Discourse Grammar.
259 pages. 2005. ISBN 3-03910-696-1 · US-ISBN 0-8204-7558-0

Vol. 27 Debbie G. E. Ho
Classroom Talk. Exploring the Sociocultural Structure of Formal ESL Learning.
2nd edition. 254 pages. 2006, 2007. ISBN 978-3-03911-434-4

Vol. 28 Javier Pérez-Guerra, Dolores González-Álvarez, Jorge L. Bueno-Alonso & Esperanza Rama-Martínez (eds)
'Of Varying Language and Opposing Creed'. New Insights into Late Modern English.
455 pages. 2007. ISBN 978-3-03910-788-9

Vol. 29 Francesca Bargiela-Chiappini & Maurizio Gotti (eds)
Asian Business Discourse(s).
350 pages. 2005. ISBN 3-03910-804-2 · US-ISBN 0-8204-7574-2

Vol. 30 Nicholas Brownlees (ed.)
News Discourse in Early Modern Britain. Selected Papers of CHINED 2004.
300 pages. 2006. ISBN 3-03910-805-0 · US-ISBN 0-8204-8025-8

Vol. 31 Roberta Facchinetti & Matti Rissanen (eds)
Corpus-based Studies of Diachronic English.
300 pages. 2006. ISBN 3-03910-851-4 · US-ISBN 0-8204-8040-1

Vol. 32 Marina Dossena & Susan M. Fitzmaurice (eds)
Business and Official Correspondence. Historical Investigations.
209 pages. 2006. ISBN 3-03910-880-8 · US-ISBN 0-8204-8352-4

Vol. 33 Giuliana Garzone & Srikant Sarangi (eds)
Discourse, Ideology and Specialized Communication.
494 pages. 2007. ISBN 978-3-03910-888-6

Vol. 34 Giuliana Garzone & Cornelia Ilie (eds)
The Use of English in Institutional and Business Settings.
An Intercultural Perspective.
372 pages. 2007. ISBN 978-3-03910-889-3

Vol. 35 Vijay K. Bhatia & Maurizio Gotti (eds)
Explorations in Specialized Genres.
316 pages. 2006. ISBN 3-03910-995-2 · US-ISBN 0-8204-8372-9

Vol. 36 Heribert Picht (ed.)
Modern Approaches to Terminological Theories and Applications.
432 pages. 2006. ISBN 3-03911-156-6 · US-ISBN 0-8204-8380-X

Vol. 37 Anne Wagner & Sophie Cacciaguidi-Fahy (eds)
Legal Language and the Search for Clarity / Le langage juridique et la quête de clarté.
Practice and Tools / Pratiques et instruments.
487 pages. 2006. ISBN 3-03911-169-8 · US-ISBN 0-8204-8388-5

Vol. 38 Juan Carlos Palmer-Silveira, Miguel F. Ruiz-Garrido &
Inmaculada Fortanet-Gómez (eds)
Intercultural and International Business Communication.
Theory, Research and Teaching.
2nd edition. 343 pages. 2006, 2008. ISBN 978-3-03911-680-5

Vol. 39 Christiane Dalton-Puffer, Dieter Kastovsky, Nikolaus Ritt &
Herbert Schendl (eds)
Syntax, Style and Grammatical Norms. English from 1500–2000.
250 pages. 2006. ISBN 3-03911-181-7 · US-ISBN 0-8204-8394-X

Vol. 40 Marina Dossena & Irma Taavitsainen (eds)
Diachronic Perspectives on Domain-Specific English.
280 pages. 2006. ISBN 3-03910-176-0 · US-ISBN 0-8204-8391-5

Vol. 41 John Flowerdew & Maurizio Gotti (eds)
Studies in Specialized Discourse.
293 pages. 2006. ISBN 3-03911-178-7

Vol. 42 Ken Hyland & Marina Bondi (eds)
Academic Discourse Across Disciplines.
320 pages. 2006. ISBN 3-03911-183-3 · US-ISBN 0-8204-8396-6

Vol. 43 Paul Gillaerts & Philip Shaw (eds)
The Map and the Landscape. Norms and Practices in Genre.
256 pages. 2006. ISBN 3-03911-182-5 · US-ISBN 0-8204-8395-4

Vol. 44 Maurizio Gotti & Davide Giannoni (eds)
New Trends in Specialized Discourse Analysis.
301 pages. 2006. ISBN 3-03911-184-1 · US-ISBN 0-8204-8381-8

Vol. 45 Maurizio Gotti & Françoise Salager-Meyer (eds)
Advances in Medical Discourse Analysis. Oral and Written Contexts.
492 pages. 2006. ISBN 3-03911-185-X · US-ISBN 0-8204-8382-6

Vol. 46 Maurizio Gotti & Susan Šarcević (eds)
Insights into Specialized Translation.
396 pages. 2006. ISBN 3-03911-186-8 · US-ISBN 0-8204-8383-4

Vol. 47 Khurshid Ahmad & Margaret Rogers (eds)
Evidence-based LSP. Translation, Text and Terminology.
584 pages. 2007. ISBN 978-3-03911-187-9

Vol. 48 Hao Sun & Dániel Z. Kádár (eds)
It's the Dragon's Turn. Chinese Institutional Discourses.
262 pages. 2008. ISBN 978-3-03911-175-6

Vol. 49 Cristina Suárez-Gómez
Relativization in Early English (950-1250). the Position of Relative Clauses.
149 pages. 2006. ISBN 3-03911-203-1 · US-ISBN 0-8204-8904-2

Vol. 50 Maria Vittoria Calvi & Luisa Chierichetti (eds)
Nuevas tendencias en el discurso de especialidad.
319 pages. 2006. ISBN 978-3-03911-261-6

Vol. 51 Mari Carmen Campoy & María José Luzón (eds)
Spoken Corpora in Applied Linguistics.
274 pages. 2008. ISBN 978-3-03911-275-3

Vol. 52 Konrad Ehlich & Dorothee Heller (Hrsg.)
Die Wissenschaft und ihre Sprachen.
323 pages. 2006. ISBN 978-3-03911-272-2

Vol. 53 Jingyu Zhang
The Semantic Salience Hierarchy Model. The L2 Acquisition of Psych Predicates
273 pages. 2007. ISBN 978-3-03911-300-2

Vol. 54 Norman Fairclough, Giuseppina Cortese & Patrizia Ardizzone (eds)
Discourse and Contemporary Social Change.
555 pages. 2007. ISBN 978-3-03911-276-0

Vol. 55 Jan Engberg, Marianne Grove Ditlevsen, Peter Kastberg & Martin Stegu (eds)
New Directions in LSP Teaching.
331 pages. 2007. ISBN 978-3-03911-433-7

Vol. 56 Dorothee Heller & Konrad Ehlich (Hrsg.)
Studien zur Rechtskommunikation.
322 pages. 2007. ISBN 978-3-03911-436-8

Vol. 57 Teruhiro Ishiguro & Kang-kwong Luke (eds)
Grammar in Cross-Linguistic Perspective.
The Syntax, Semantics, and Pragmatics of Japanese and Chinese.
304 pages. 2012. ISBN 978-3-03911-445-0

Vol. 58 Carmen Frehner
Email – SMS – MMS
294 pages. 2008. ISBN 978-3-03911-451-1

Vol. 59 Isabel Balteiro
The Directionality of Conversion in English. A Dia-Synchronic Study.
276 pages. 2007. ISBN 978-3-03911-241-8

Vol. 60 Maria Milagros Del Saz Rubio
English Discourse Markers of Reformulation.
237 pages. 2007. ISBN 978-3-03911-196-1

Vol. 61 Sally Burgess & Pedro Martín-Martín (eds)
English as an Additional Language in Research Publication and Communication.
259 pages. 2008. ISBN 978-3-03911-462-7

Vol. 62 Sandrine Onillon
Pratiques et représentations de l'écrit.
458 pages. 2008. ISBN 978-3-03911-464-1

Vol. 63 Hugo Bowles & Paul Seedhouse (eds)
Conversation Analysis and Language for Specific Purposes.
2nd edition. 337 pages. 2007, 2009. ISBN 978-3-0343-0045-2

Vol. 64 Vijay K. Bhatia, Christopher N. Candlin & Paola Evangelisti Allori (eds)
Language, Culture and the Law.
The Formulation of Legal Concepts across Systems and Cultures.
342 pages. 2008. ISBN 978-3-03911-470-2

Vol. 65 Jonathan Culpeper & Dániel Z. Kádár (eds)
Historical (Im)politeness.
300 pages. 2010. ISBN 978-3-03911-496-2

Vol. 66 Linda Lombardo (ed.)
Using Corpora to Learn about Language and Discourse.
237 pages. 2009. ISBN 978-3-03911-522-8

Vol. 67 Natsumi Wakamoto
Extroversion/Introversion in Foreign Language Learning.
Interactions with Learner Strategy Use.
159 pages. 2009. ISBN 978-3-03911-596-9

Vol. 68 Eva Alcón-Soler (ed.)
Learning How to Request in an Instructed Language Learning Context.
260 pages. 2008. ISBN 978-3-03911-601-0

Vol. 69 Domenico Pezzini
The Translation of Religious Texts in the Middle Ages.
428 pages. 2008. ISBN 978-3-03911-600-3

Vol. 70 Tomoko Tode
Effects of Frequency in Classroom Second Language Learning.
Quasi-experiment and stimulated-recall analysis.
195 pages. 2008. ISBN 978-3-03911-602-7

Vol. 71 Egor Tsedryk
Fusion symétrique et alternances ditransitives.
211 pages. 2009. ISBN 978-3-03911-609-6

Vol. 72 Cynthia J. Kellett Bidoli & Elana Ochse (eds)
English in International Deaf Communication.
444 pages. 2008. ISBN 978-3-03911-610-2

Vol. 73 Joan C. Beal, Carmela Nocera & Massimo Sturiale (eds)
Perspectives on Prescriptivism.
269 pages. 2008. ISBN 978-3-03911-632-4

Vol. 74 Carol Taylor Torsello, Katherine Ackerley & Erik Castello (eds)
Corpora for University Language Teachers.
308 pages. 2008. ISBN 978-3-03911-639-3

Vol. 75 María Luisa Pérez Cañado (ed.)
English Language Teaching in the European Credit Transfer System. Facing the Challenge.
251 pages. 2009. ISBN 978-3-03911-654-6

Vol. 76 Marina Dossena & Ingrid Tieken-Boon van Ostade (eds)
Studies in Late Modern English Correspondence. Methodology and Data.
291 pages. 2008. ISBN 978-3-03911-658-4

Vol. 77 Ingrid Tieken-Boon van Ostade & Wim van der Wurff (eds)
Current Issues in Late Modern English.
436 pages. 2009. ISBN 978-3-03911-660-7

Vol. 78 Marta Navarro Coy (ed.)
Practical Approaches to Foreign Language Teaching and Learning.
297 pages. 2009. ISBN 978-3-03911-661-4

Vol. 79 Qing Ma
Second Language Vocabulary Acquisition.
333 pages. 2009. ISBN 978-3-03911-666-9

Vol. 80 Martin Solly, Michelangelo Conoscenti & Sandra Campagna (eds)
Verbal/Visual Narrative Texts in Higher Education.
384 pages. 2008. ISBN 978-3-03911-672-0

Vol. 81 Meiko Matsumoto
From Simple Verbs to Periphrastic Expressions: The Historical Development of Composite Predicates, Phrasal Verbs, and Related Constructions in English.
235 pages. 2008. ISBN 978-3-03911-675-1

Vol. 82 Melinda Dooly
Doing Diversity. Teachers' Construction of Their Classroom Reality.
180 pages. 2009. ISBN 978-3-03911-687-4

Vol. 83 Victoria Guillén-Nieto, Carmen Marimón-Llorca & Chelo Vargas-Sierra (eds)
Intercultural Business Communication and Simulation and Gaming Methodology.
392 pages. 2009. ISBN 978-3-03911-688-1

Vol. 84 Maria Grazia Guido
English as a Lingua Franca in Cross-cultural Immigration Domains.
285 pages. 2008. ISBN 978-3-03911-689-8

Vol. 85 Erik Castello
Text Complexity and Reading Comprehension Tests.
352 pages. 2008. ISBN 978-3-03911-717-8

Vol. 86 Maria-Lluisa Gea-Valor, Isabel García-Izquierdo & Maria-José Esteve (eds)
Linguistic and Translation Studies in Scientific Communication.
317 pages. 2010. ISBN 978-3-0343-0069-8

Vol. 87 Carmen Navarro, Rosa Mª Rodríguez Abella, Francesca Dalle Pezze & Renzo Miotti (eds)
La comunicación especializada.
355 pages. 2008. ISBN 978-3-03911-733-8

Vol. 88 Kiriko Sato
The Development from Case-Forms to Prepositional Constructions in Old English Prose.
231 pages. 2009. ISBN 978-3-03911-763-5

Vol. 89 Dorothee Heller (Hrsg.)
Formulierungsmuster in deutscher und italienischer Fachkommunikation. Intra- und interlinguale Perspektiven.
315 pages. 2008. ISBN 978-3-03911-778-9

Vol. 90 Henning Bergenholtz, Sandro Nielsen & Sven Tarp (eds)
Lexicography at a Crossroads. Dictionaries and Encyclopedias Today, Lexicographical Tools Tomorrow.
372 pages. 2009. ISBN 978-3-03911-799-4

Vol. 91 Manouchehr Moshtagh Khorasani
The Development of Controversies. From the Early Modern Period to Online Discussion Forums.
317 pages. 2009. ISBN 978-3-3911-711-6

Vol. 92 María Luisa Carrió-Pastor (ed.)
Content and Language Integrated Learning. Cultural Diversity.
178 pages. 2009. ISBN 978-3-3911-818-2

Vol. 93 Roger Berry
Terminology in English Language Teaching. Nature and Use.
262 pages. 2010. ISBN 978-3-0343-0013-1

Vol. 94 Roberto Cagliero & Jennifer Jenkins (eds)
Discourses, Communities, and Global Englishes
240 pages. 2010. ISBN 978-3-0343-0012-4

Vol. 95 Facchinetti Roberta, Crystal David, Seidlhofer Barbara (eds)
From International to Local English – And Back Again.
268 pages. 2010. ISBN 978-3-0343-0011-7

Vol. 96 Cesare Gagliardi & Alan Maley (eds)
EIL, ELF, Global English. Teaching and Learning Issues
376 pages. 2010. ISBN 978-3-0343-0010-0

Vol. 97 Sylvie Hancil (ed.)
The Role of Prosody in Affective Speech.
403 pages. 2009. ISBN 978-3-03911-696-6

Vol. 98 Marina Dossena & Roger Lass (eds)
Studies in English and European Historical Dialectology.
257 pages. 2009. ISBN 978-3-0343-0024-7

Vol. 99 Christine Béal
Les interactions quotidiennes en français et en anglais. De l'approche comparative à l'analyse des situations interculturelles.
424 pages. 2010. ISBN 978-3-0343-0027-8

Vol. 100 Maurizio Gotti (ed.)
Commonality and Individuality in Academic Discourse.
398 pages. 2009. ISBN 978-3-0343-0023-0

Vol. 101 Javier E. Díaz Vera & Rosario Caballero (eds)
Textual Healing. Studies in Medieval English Medical, Scientific and Technical Texts.
213 pages. 2009. ISBN 978-3-03911-822-9

Vol. 102 Nuria Edo Marzá
The Specialised Lexicographical Approach. A Step further in Dictionary-making.
316 pages. 2009. ISBN 978-3-0343-0043-8

Vol. 103 Carlos Prado-Alonso, Lidia Gómez-García, Iria Pastor-Gómez & David Tizón-Couto (eds)
New Trends and Methodologies in Applied English Language Research. Diachronic, Diatopic and Contrastive Studies.
348 pages. 2009. ISBN 978-3-0343-0046-9

Vol. 104 Françoise Salager-Meyer & Beverly A. Lewin
Crossed Words. Criticism in Scholarly Writing?
371 pages. 2011. ISBN 978-3-0343-0049-0.

Vol. 105 Javier Ruano-García
Early Modern Northern English Lexis. A Literary Corpus-Based Study.
611 pages. 2010. ISBN 978-3-0343-0058-2

Vol. 106 Rafael Monroy-Casas
Systems for the Phonetic Transcription of English. Theory and Texts.
280 pages. 2011. ISBN 978-3-0343-0059-9

Vol. 107 Nicola T. Owtram
The Pragmatics of Academic Writing.
A Relevance Approach to the Analysis of Research Article Introductions.
311 pages. 2009. ISBN 978-3-0343-0060-5

Vol. 108 Yolanda Ruiz de Zarobe, Juan Manuel Sierra & Francisco Gallardo del Puerto (eds)
Content and Foreign Language Integrated Learning.
Contributions to Multilingualism in European Contexts
343 pages. 2011. ISBN 978-3-0343-0074-2

Vol. 109 Ángeles Linde López & Rosalía Crespo Jiménez (eds)
Professional English in the European context. The EHEA challenge.
374 pages. 2010. ISBN 978-3-0343-0088-9

Vol. 110 Rosalía Rodríguez-Vázquez
The Rhythm of Speech, Verse and Vocal Music. A New Theory.
394 pages. 2010. ISBN 978-3-0343-0309-5

Vol. 111 Anastasios Tsangalidis & Roberta Facchinetti (eds)
Studies on English Modality. In Honour of Frank Palmer.
392 pages. 2009. ISBN 978-3-0343-0310-1

Vol. 112 Forthcoming.

Vol. 113 Mihhail Lotman & Maria-Kristiina Lotman (eds)
Frontiers in Comparative Prosody. In memoriam: Mikhail Gasparov.
426 pages. 2011. ISBN 978-3-0343-0373-6

Vol. 114 Merja Kytö, John Scahill & Harumi Tanabe (eds)
Language Change and Variation from Old English to Late Modern English.
A Festschrift for Minoji Akimoto
422 pages. 2010. ISBN 978-3-0343-0372-9

Vol. 115 Giuliana Garzone & Paola Catenaccio (eds)
Identities across Media and Modes. Discursive Perspectives.
379 pages. 2009. ISBN 978-3-0343-0386-6

Vol. 116 Elena Landone
Los marcadores del discurso y cortesía verbal en español.
390 pages. 2010. ISBN 978-3-0343-0413-9

Vol. 117 Maurizio Gotti & Christopher Williams (eds)
Legal Discourse across Languages and Cultures.
339 pages. 2010. ISBN 978-3-0343-0425-2

Vol. 118 David Hirsh
Academic Vocabulary in Context.
217 pages. 2010. ISBN 978-3-0343-0426-9

Vol. 119 Yvonne Dröschel
Lingua Franca English. The Role of Simplification and Transfer.
358 pages. 2011. ISBN 978-3-0343-0432-0

Vol. 120 Tengku Sepora Tengku Mahadi, Helia Vaezian & Mahmoud Akbari
Corpora in Translation. A Practical Guide.
135 pages. 2010. ISBN 978-3-0343-0434-4

Vol. 121 Davide Simone Giannoni & Celina Frade (eds)
Researching Language and the Law. Textual Features and Translation Issues.
278 pages. 2010. ISBN 978-3-0343-0443-6

Vol. 122 Daniel Madrid & Stephen Hughes (eds)
Studies in Bilingual Education.
472 pages. 2011. ISBN 978-3-0343-0474-0

Vol. 123 Vijay K. Bhatia, Christopher N. Candlin & Maurizio Gotti (eds)
The Discourses of Dispute Resolution.
290 pages. 2010. ISBN 978-3-0343-0476-4

Vol. 124 Davide Simone Giannoni
Mapping Academic Values in the Disciplines. A Corpus-Based Approach.
288 pages. 2010. ISBN 978-3-0343-0488-7

Vol. 125 Giuliana Garzone & James Archibald (eds)
Discourse, Identities and Roles in Specialized Communication.
419 pages. 2010. ISBN 978-3-0343-0494-8

Vol. 126 Iria Pastor-Gómez
The Status and Development of N+N Sequences in
Contemporary English Noun Phrases.
216 pages. 2011. ISBN 978-3-0343-0534-1

Vol. 127 Carlos Prado-Alonso
Full-verb Inversion in Written and Spoken English.
261 pages. 2011. ISBN 978-3-0343-0535-8

Vol. 128 Tony Harris & María Moreno Jaén (eds)
Corpus Linguistics in Language Teaching.
214 pages. 2010. ISBN 978-3-0343-0524-2

Vol. 129 Tetsuji Oda & Hiroyuki Eto (eds)
Multiple Perspectives on English Philology and History of Linguistics.
A Festschrift for Shoichi Watanabe on his 80th Birthday.
378 pages. 2010. ISBN 978-3-0343-0480-1

Vol. 130 Luisa Chierichetti & Giovanni Garofalo (eds)
Lengua y Derecho. líneas de investigación interdisciplinaria.
283 pages. 2010. 978-3-0343-0463-4

Vol. 131 Paola Evangelisti Allori & Giuliana Garzone (eds)
Discourse, Identities and Genres in Corporate Communication.
Sponsorship, Advertising and Organizational Communication.
324 pages. 2011. 978-3-0343-0591-4

Vol. 132 Leyre Ruiz de Zarobe & Yolanda Ruiz de Zarobe (eds)
Speech Acts and Politeness across Languages and Cultures.
402 pages. 2012. 978-3-0343-0611-9

Vol. 133 Thomas Christiansen
Cohesion. A Discourse Perspective.
387 pages. 2011. 978-3-0343-0619-5

Vol. 134 Giuliana Garzone & Maurizio Gotti
Discourse, Communication and the Enterprise. Genres and Trends.
451 pages. 2011. ISBN 978-3-0343-0620-1

Vol. 135 Zsuzsa Hoffmann
Ways of the World's Words.
Language Contact in the Age of Globalization.
334 pages 2011. ISBN 978-3-0343-0673-7

Vol. 136 Cecilia Varcasia (ed.)
Becoming Multilingual.
Language Learning and Language Policy between Attitudes and Identities.
213 pages. 2011. ISBN 978-3-0343-0687-5

Vol. 137 Susy Macqueen
The Emergence of Patterns in Second Language Writing.
A Sociocognitive Exploration of Lexical Trails.
325 pages. 2012. ISBN 978-3-0343-1010-9

Vol. 138 Maria Vittoria Calvi & Giovanna Mapelli (eds)
La lengua del turismo. Géneros discursivos y terminología.
365 pages. 2011. ISBN 978-3-0343-1011-6

Vol. 139 Ken Lau
Learning to Become a Professional in a Textually-Mediated World.
A Text-Oriented Study of Placement Practices.
261 pages. 2012. ISBN 978-3-0343-1016-1

Vol. 140 Sandra Campagna, Giuliana Garzone, Cornelia Ilie & Elizabeth Rowley-Jolivet (eds)
Evolving Genres in Web-mediated Communication.
337 pages. 2012. ISBN 978-3-0343-1013-0

Vol. 141 Edith Esch & Martin Solly (eds)
The Sociolinguistics of Language Education in International Contexts.
263 pages. 2012. ISBN 978-3-0343-1009-3

Vol. 142 Forthcoming.

Vol. 143 David Tizón-Couto
Left Dislocation in English. A Functional-Discoursal Approach.
416 pages. 2012. ISBN 978-3-0343-1037-6

Vol. 144 Margrethe Petersen & Jan Engberg (eds)
Current Trends in LSP Research. Aims and Methods.
323 pages. 2011. ISBN 978-3-0343-1054-3

Vol. 145 David Tizón-Couto, Beatriz Tizón-Couto, Iria Pastor-Gómez & Paula Rodríguez-Puente (eds)
New Trends and Methodologies in Applied English Language Research II.
Studies in Language Variation, Meaning and Learning.
283 pages. 2012. ISBN 978-3-0343-1061-1

Vol. 146 Rita Salvi & Hiromasa Tanaka (eds)
Intercultural Interactions in Business and Management.
306 pages. 2011. ISBN 978-3-0343-1039-0

Vol. 147 Francesco Straniero Sergio & Caterina Falbo (eds)
Breaking Ground in Corpus-based Interpreting Studies.
254 pages. 2012. ISBN 978-3-0343-1071-0

Vol. 148 Forthcoming.

Vol. 149 Vijay K. Bhatia & Paola Evangelisti Allori (eds)
Discourse and Identity in the Professions. Legal, Corporate and Institutional Citizenship.
352 pages. 2011. ISBN 978-3-0343-1079-6

Vol. 150 Maurizio Gotti (ed.)
Academic Identity Traits. A Corpus-Based Investigation.
363 pages. 2012. ISBN 978-3-0343-1141-0

Vol. 151 Priscilla Heynderickx, Sylvain Dieltjens, Geert Jacobs, Paul Gillaerts & Elizabeth de Groot (eds)
The Language Factor in International Business.
New Perspectives on Research, Teaching and Practice.
320 pages. 2012. ISBN 978-3-0343-1090-1

Vol. 152 Paul Gillaerts, Elizabeth de Groot, Sylvain Dieltjens, Priscilla Heynderickx & Geert Jacobs (eds)
Researching Discourse in Business Genres. Cases and Corpora.
215 pages. 2012. ISBN 978-3-0343-1092-5

Vol. 153 Yongyan Zheng
Dynamic Vocabulary Development in a Foreign Language.
262 pages. 2012. ISBN 978-3-0343-1106-9

Vol. 154 Carmen Argondizzo (ed.)
Creativity and Innovation in Language Education.
357 pages. 2012. ISBN 978-3-0343-1080-2

Vol. 155 David Hirsh (ed.)
Current Perspectives in Second Language Vocabulary Research.
180 pages. 2012. ISBN 978-3-0343-1108-3

Vol. 156 Seiji Shinkawa
Unhistorical Gender Assignment in Laȝamon's *Brut*. A Case Study of a Late Stage in the Development of Grammatical Gender toward its Ultimate Loss.
186 pages. 2012. ISBN 978-3-0343-1124-3

Vol. 157 Forthcoming.

Vol. 158 Bárbara Eizaga Rebollar (ed.)
Studies in Linguistics and Cognition.
301 pages. 2012. ISBN 978-3-0343-1138-0

Vol. 159 Giuliana Garzone, Paola Catenaccio, Chiara Degano (eds)
Genre Change in the Contemporary World. Short-term Diachronic Perspectives.
329 pages. 2012. ISBN 978-3-0343-1214-1

Vol. 160 Carol Berkenkotter, Vijay K. Bhatia & Maurizio Gotti (eds)
Insights into Academic Genres.
468 pages. 2012. ISBN 978-3-0343-1211-0

Vol. 161 Forthcoming.

Vol. 162 Patrizia Anesa
Jury Trials and the Popularization of Legal Language. A Discourse Analytical Approach.
247 pages. 2012. ISBN 978-3-0343-1231-8

Vol. 163-165 Forthcoming.

Vol. 166 Rita Salvi & Janet Bowker (eds)
Space, Time and the Construction of Identity.
Discursive Indexicality in Cultural, Institutional and Professional Fields.
324 pages. 2013. ISBN 978-3-0343-1254-7

Vol. 167 Shunji Yamazaki & Robert Sigley (eds)
Approaching Language Variation through Corpora. A Festschrift in Honour of Toshio Saito.
421 pages. 2013. ISBN 978-3-0343-1264-6

Vol. 168 Franca Poppi
Global Interactions in English as a Lingua Franca. How written communication is changing under the influence of electronic media and new contexts of use.
249 pages. 2012. ISBN 978-3-0343-1276-9

Vol. 169-170 Forthcoming.

Vol. 171 Aleksandra Matulewska
Legilinguistic Translatology. A Parametric Approach to Legal Translation.
279 pages. 2013. ISBN 978-3-0343-1287-5

Vol. 172 Maurizio Gotti & Carmen Sancho Guinda (eds)
Narratives in Academic and Professional Genres.
513 pages. 2013. ISBN 978-3-0343-1371-1

www.ingramcontent.com/pod-product-compliance
Ingram Content Group UK Ltd.
Pitfield, Milton Keynes, MK11 3LW, UK
UKHW021852100926
8556IPUK00020B/1195

9 783034 312547